Passive Intermodulation

Space Science, Technology and Application Series

Series Editors
Weimin Bao, Yulin Deng

**Simulation Method of Multipactor and Its Application in
Satellite Microwave Components**
Wanzhao Cui, Yun Li, Hongtai Zhang, Jing Yang

Passive Intermodulation
Concepts and Technology
Edited by Wanzhao Cui, Jun Li, Huan Wei, Xiang Chen

For more information about this series, please visit: https://www.routledge.com/Space-Science-Technology-and-Application-Series/book-series/SSTA

Passive Intermodulation

Concepts and Technology

Edited by

Wanzhao Cui,
Jun Li, Huan Wei, and
Xiang Chen

CRC Press
Taylor & Francis Group
Boca Raton London New York

CRC Press is an imprint of the
Taylor & Francis Group, an **informa** business

北京理工大学出版社
BEIJING INSTITUTE OF TECHNOLOGY PRESS

First edition published 2022
by CRC Press
6000 Broken Sound Parkway NW, Suite 300, Boca Raton, FL 33487-2742

and by CRC Press
2 Park Square, Milton Park, Abingdon, Oxon, OX14 4RN

CRC Press is an imprint of Taylor & Francis Group, LLC

© 2022 selection and editorial matter, Wanzhao Cui, Jun Li, Huan Wei, Xiang Chen; individual chapters, the contributors

Library of Congress Cataloging-in-Publication Data
Names: Cui, Wanzhao, 1975- editor.
Title: Passive intermodulation : concepts and technology / edited by
Cui Wanzhao, Li Jun, Huan Wei, Chen Xiang.
Description: First edition. | Boca Raton : CRC Press, [2022] |
Series: Space science, technology and application series | Includes
bibliographical references and index. |
Summary: "Focusing on Passive Intermodulation (PIM), a common instance of distortion in high-power multi-channel communication systems, the book introduces the concepts and Technology of PIM. In the context of "space-ground-object" integrated communication and increasing spectrum resource constraints, the PIM problem is becoming increasingly prominent, while microwave components and system-level PIM are facing new challenges. Based on the experience and achievements in engineering practice of the China Academy of Space Technology in Xi'an, this title describes the basic theory, generation mechanism, analysis and evaluation methods, location detection and suppression techniques for microwave components of spacecraft PIM from a theoretical and engineering perspective. At the same time, the latest achievements of microwave components PIM are outlined in a certain technical depth, along with its operational guidance and paths for further advancement. The book will be a useful reference for researchers, engineering and students interested in microwave technology, satellite technology and mobile communication technology"—Provided by publisher.
Identifiers: LCCN 2021046955 (print) | LCCN 2021046956 (ebook) |
ISBN 9781032217604 (hbk) | ISBN 9781032217673 (pbk) | ISBN 9781003269953 (ebk)
Subjects: LCSH: Electromagnetic interference. | Microwave communication systems. |
Electric distortion. | Signal integrity (Electronics)
Classification: LCC TK7867.2 .P37 2022 (print) | LCC TK7867.2 (ebook) |
DDC 621.382/24—dc23/eng/20211110
LC record available at https://lccn.loc.gov/2021046955
LC ebook record available at https://lccn.loc.gov/2021046956

ISBN: 978-1-032-21760-4 (hbk)
ISBN: 978-1-032-21767-3 (pbk)
ISBN: 978-1-003-26995-3 (ebk)

DOI: 10.1201/9781003269953

Typeset in Minion
by codeMantra

Contents

Contents

Foreword

WITH THE RAPID DEVELOPMENT of satellite payload technology, higher requirements for transmission rate and transmission distance are put forward, and the satellite communication system is developing toward high power, multicarrier, broadband and high sensitivity. Considering the miniaturization and lightweight demand of satellite payloads, its system tends to adopt a common antenna structure for transceiver. With the increasingly intensive use of wireless communication spectrum, the gradual application of high-power transmitters, the increasing sensitivity of receivers and the increasing popularity of colocation phenomenon, passive intermodulation caused by the weak nonlinearity of passive components can no longer be ignored and even becomes one of the main interferences of wireless communication systems; especially with the continuous development of mobile communication 5G and beyond 5G, the increasing congestion of spectrum resources and the continuous overlapping of frequency bands have become more serious. Obviously, passive intermodulation has become a common problem faced by satellite and ground base stations, which will have an important impact on the future integrated communication construction of satellite and ground base stations.

For engineering designers, passive intermodulation should belong to the scope of physical research, which involves the basic research content in the fields of material properties, processing technology and structure design; for natural science researchers, passive intermodulation should be classified as an engineering phenomenon, which will affect directly the ratio of signal-to-noise in the receiver once it enters the receiving channel, and the severe nonlinearity will overwhelm directly the received signal, resulting in a complete loss of communication. Due to the interdisciplinary nature of passive intermodulation and its importance, passive intermodulation has become a hot research topic in several fields in recent years.

The problem of passive intermodulation has developed over a period of nearly half a century from the emergence of the 43rd-order passive intermodulation of the European maritime satellite MARECS in the 1970s to the current rapid development of air-space mobile communications, which has gradually become a necessary performance criterion for the assessment of microwave components of communication systems. At present, engineering researchers have a preliminary understanding of the passive intermodulation sources, for example, by avoiding the use of ferromagnetic materials in microwave components to suppress passive intermodulation, but further suppression of passive intermodulation will require the analysis of its sources. China's communication satellites have encountered the problem of passive intermodulation many times during the development process, and it even once became the core bottleneck restricting the development of technology. Facing the actual problems of major space projects, we have carried out a series of targeted research. This book is a systematic and novel summary of these research works of the groups for many years, including a detailed discussion of the generation mechanism of passive intermodulation, passive intermodulation analysis, passive intermodulation detection and localization and passive intermodulation suppression and can be a good reference for engineering designers and basic researchers.

Lvqian Zhang
Academician of Chinese Academy of Engineering

Preface

SPACE-GROUND-SEA INTEGRATED COMMUNICATION HAS become an inevitable direction of future communication development. The sky is far away, and transmission is not easy; spacecraft are indispensable for the interconnection of everything. To ensure faster, more accurate and more extensive communication, China's new generation of spacecraft payload technology is developing toward the direction of high-power, high-performance and high reliability. Therefore, the power of microwave components is increasing, which puts forward increased requirements for its antipassive intermodulation design technology. Once the passive intermodulation interference causes serious consequences, the noise level will be increased, so that the system cannot work properly. To trace the source of passive intermodulation, the mechanism of passive intermodulation microwave components must be analyzed deeply, the specific location of passive intermodulation generated shall be determined, the passive intermodulation power level needs to be measured, and the design of microwave components antipassive intermodulation must be completed, and thus, the technical level of spacecraft can effectively be improved.

This book consists of six chapters which are completed by Wanzhao Cui, Jun Li, Huan Wei and Xiang Chen et al. Wanzhao Cui and Jun Li are responsible for the structural design of the book, and Huan Wei completed the unified draft of the book. In writing, Huan Wei was responsible for Chapter 1, Chunjiang Bai for Chapter 2, Rui Wang for Chapter 3, Xiang Chen for Chapters 4 and 5, and Tiancun Hu for Chapter 6. In addition, Na Zhang was involved in writing Chapter 1, Wanzhao Cui was involved in writing Chapter 4, Jun Li was involved in writing Chapter 5, Yun He was involved in writing Section 2.6, He Bai was involved in writing Sections 3.5, 6.6 and 6.7, and Qi Wang was involved in writing Sections 6.4 and 6.5.

Lvqian Zhang, academician of the Chinese Academy of Engineering, wrote the Foreword for this book and offered many valuable comments during his busy schedule. The research work of this book was supported by the Key Project of National Natural Science Foundation of China (U1537211), the National Natural Science Foundation of China (11705142, 11605135, 61701394, 61801376, 5165421, 11675278, 51827809, 61901359) and the Key Laboratory of Space Microwave Technology Stabilization Fund projects (HTKJ2020KL504002, HTKJ2019KL504009). The publication of this book was supported by the National Publication Fund and was awarded the "13th 5-Year Plan" National Key Publication Project and the "National Pillar" Publication Project. The book has also received support from Vice President Li, Chief Engineer Hongxi Yu, Deputy Chief Engineer

Zhengjun Li and Deputy Director Xiaojun Li of China Academy of Space Technology (Xi'an). The team members Xinbo Wang, Yun Li, Guibai Xie, Jing Yang, Qiangqiang Song, Xiaoxiao Li and Guobao Feng also participated in the passive intermodulation research. We thank Yongning He, Jianfeng Zhang, Lixin Ran, Jiangtao Huangfu, Jianping An, Xiangyuan Bu, Yongjun Xie, Tuanjie Li and Wanshun Jiang for their great work on the passive intermodulation research, and special thanks to Xiaolong Zhao, Songchang Zhang, Xiong Chen, Lei Zhang, Dongwei Wu, Yuru Mao, Mei Zhang, Chuan Zheng, Lu Tian, Xiaozheng Gao and Bizheng Liang for their work during their PhD/MS degrees. This book has received great support from Beijing Institute of Technology Press during the publication process. In particular, editors Bingquan Li, Haili Zhang, Shu Sun, Xian Zeng and Shan Guo have done a lot of work for the publication of this book, and we would like to thank them all.

With the rapid development of science and technology, technological progress will never stop. Although the authors have exhausted their research results in writing the whole book, they have not been able to fully cover all aspects of passive intermodulation mechanism, analysis, detection and suppression. With the improvement of data processing technology and process technology, passive intermodulation detection and suppression will definitely achieve greater development. Although we have tried our best, the manuscript inevitably has some omissions and deficiencies due to our level and ability, and we sincerely invite readers and experts to criticize and correct our work.

Editors

Wanzhao Cui is a professor at the National Key Laboratory of Science and Technology on Space Microwave, at the China Academy of Space Technology, Xi'an, China. His current research interests include microwave technology and satellite communication.

Jun Li is a professor at the National Key Laboratory of Science and Technology on Space Microwave, at the China Academy of Space Technology, Xi'an, China. He is engaged in research on microwave technology and satellite communication.

Huan Wei is a senior engineer at the National Key Laboratory of Science and Technology on Space Microwave, at the China Academy of Space Technology, Xi'an, China. Her expertise is in the field of space microwave technology.

Xiang Chen is a senior engineer at the National Key Laboratory of Science and Technology on Space Microwave, at the China Academy of Space Technology, Xi'an, China. His research interests are microwave and millimeter-wave technology.

Contributors

Jianping An
Beijing Institute of Technology
China

Chunjiang Bai
National Key Laboratory of Science and
    Technology on Space Microwave
China Academy of Space Technology
Xi'an, China

He Bai
National Key Laboratory of Science and
    Technology on Space Microwave
China Academy of Space Technology
Xi'an, China

Xiangyuan Bu
Beijing Institute of Technology
Beijing, China

Wanzhao Cui
National Key Laboratory of Science and
    Technology on Space Microwave
China Academy of Space Technology
Xi'an, China

Xiang Chen
National Key Laboratory of Science and
    Technology on Space Microwave
China Academy of Space Technology
Xi'an, China

Jiangtao Huangfu
Zhejiang University
Zhejiang, China

Tiancun Hu
National Key Laboratory of Science and
    Technology on Space Microwave
China Academy of Space Technology
Xi'an, China

Yun He
National Key Laboratory of Science and
    Technology on Space Microwave
China Academy of Space Technology
Xi'an, China

Yongning He
Xi'an Jiaotong University
Xi'an, China

Jun Li
National Key Laboratory of Science and
    Technology on Space Microwave
China Academy of Space Technology
Xi'an, China

Tuanjie Li
Xidian University
Xi'an, China

Xiaojun Li
National Key Laboratory of Science and
  Technology on Space Microwave
China Academy of Space Technology
Xi'an, China

Yun Li
National Key Laboratory of Science and
  Technology on Space Microwave
China Academy of Space Technology
Xi'an, China

Zhengjun Li
National Key Laboratory of Science and
  Technology on Space Microwave
China Academy of Space Technology
Xi'an, China

Changjun Liu
Sichuan University
Xi'an, China

Lixin Ran
Zhejiang University
Zhejiang, China

Lu Tian
Beijing Information Science and
  Technology University
China

Qi Wang
National Key Laboratory of Science and
  Technology on Space Microwave
China Academy of Space Technology
Xi'an, China

Rui Wang
National Key Laboratory of Science and
  Technology on Space Microwave
China Academy of Space Technology
Xi'an, China

Xinbo Wang
National Key Laboratory of Science and
  Technology on Space Microwave
China Academy of Space Technology
Xi'an, China

Huan Wei
National Key Laboratory of Science and
  Technology on Space Microwave
China Academy of Space Technology
Xi'an, China

Yongjun Xie
Beijing University of Aeronautics and
  Astronautics
Beijing, China

Jianfeng Zhang
Southeast University
Nanjing, China

Na Zhang
National Key Laboratory of Science and
  Technology on Space Microwave
China Academy of Space Technology
Xi'an, China

CHAPTER 1

Introduction

Huan Wei and Na Zhang

CONTENTS

1.1 GENERAL OVERVIEW

Passive intermodulation (PIM) refers to the interference phenomenon caused by inter-modulation signals falling into the receiving passband when two or more transmit carriers are input in a high-power multichannel communication system. In the high-power, multichannel system, the nonlinearity of these passive components will generate harmonics with higher frequency than the operating frequency, and the harmonics mixed with the operating frequency will generate a new set of frequencies, which will eventually generate a set of interference spectrum in the air and affect normal communication. Taking the example of transmitting two carriers, the nonlinearity of microwave passive components leads to carrier signals modulated by each other, resulting in a combination of carrier frequency products falling into the receive passband, which causes interference, and the reflected PIM on the transmit link will also have an impact, but the impact of PIM on the receiving passband will often have a serious impact on the system. Therefore, it is very important to analyze, calculate and detect PIM falling into the receiving passband. Figure 1.1a shows a simplified diagram of PIM interference to the system in the communication link, and Figure 1.1b shows the spectral distribution of typical PIM generation.

The typical characteristics of PIM are transmitting high power and receiving low power. For terrestrial mobile communications, it is usually required that the transmit power is 20 W/43 dBm and the received signal is 0.01 pW/−110 dBm, that is, the difference between

DOI: 10.1201/9781003269953-1

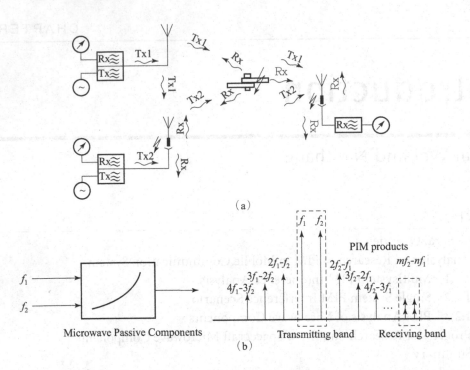

(a)

(b)

FIGURE 1.1   PIM interference: (a) simplified diagram of PIM interference to the system in the communication link; (b) spectrum distribution generated by PIM ↑—transmitted carrier signal; ↑—intermodulation interference signal; ↑—normal received signal.

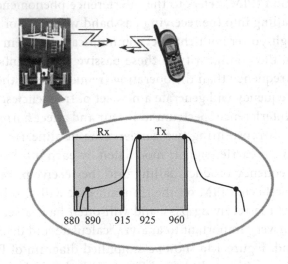

FIGURE 1.2   PIM interference in typical transceiver mobile communication system.

the transmitted signal and the received signal is 15 orders of magnitude. Typical transceiver mobile communication system PIM interference is shown in Figure 1.2.

PIM interference can occur within the same communication system, and is purely interference caused by frequency overlap within the system. Under the condition of

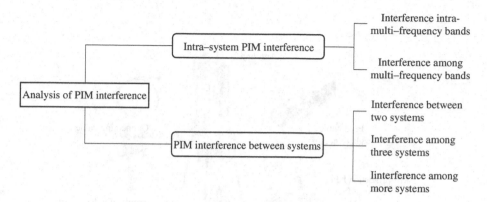

FIGURE 1.3   Typical PIM interference distribution.

multisystem common communication hardware resources, the frequencies within a single system will interfere with each other, and the spurious spectrum between systems can have an impact on the other systems; the typical interference distribution is shown in Figure 1.3. In the mobile communication system, the PIM interference caused by the nonlinearity of passive components will have a greater impact on the system performance, and the impact is more serious in the multisystem merging or adjacent site construction.

PIM interference falls into the receiving frequency band and will reduce the signal-to-noise ratio of the whole communication system. With the improvement of receiver sensitivity and transmitter power, PIM interference will affect directly the communication quality, system stability and channel capacity. With the development of mobile communication, the number of base station construction is increasing rapidly. In order to solve the problem of resource waste caused by repeated construction, there will be more and more multisystem routing scenarios.

For PIM, in the system of transmitting and receiving, the transmitting signal and the receiving signal pass through the duplexer, the high power cable, the antenna feed and the transmitting and receiving antenna at the same time. PIM-prone microwave components in typical ground mobile communication systems are shown in Figure 1.4. In this mobile communication system, if the PIM products generated by different transmitting carriers fall into the receiving band, the PIM products and the received signals will enter the receiver at the same time and form interference. However, the traditional filtering and isolation methods do not word, so the PIM problem is the key problem related to the link communication quality of the whole communication system.

1.2 ANALYSIS AND RESEARCH OF PIM IN MOBILE COMMUNICATION SYSTEM

1.2.1 Multisystem PIM Interference Analysis

Table 1.1 lists the spectrum usage of Chinese operators and railway private networks. Understanding the spectrum allocation is the basis for calculating PIM interference.

FIGURE 1.4 PIM-prone microwave components in a typical ground mobile communication system.

TABLE 1.1 Frequency Spectrum Usage of Domestic Operators and Railway Private Networks MHz

Operators	Communication Mode	Receiving Band	Transmitting Band
China Mobile	CSM900	890–909	935–954
	DCS1800	1710–1735	1805–1830
	DCS1800 (expanded)	1755–1777	1850–1872
	TD-F	1880–1920	
	TD-A	2010–2025	
	TD-E	2320–2370	
	TD-D	2570–2620	
	WLAN	2400–2483.5	
China Unicom	CSM900	909–915	954–960
	DCS1800	1745–1755	1840–1850
	WCDMA	1940–1955	2130–2145
	WLAN	2400–2483.5	
China Telecom	CDMA800	825–835	870–880
	CDMA2000	1920–1935	2110–2125
	WLAN	2400–2483.5	
China Railcom	CSM-R	885–889	930–934

TABLE 1.2   Single-System PIM Products Operated by China Mobile MHz

Communication Mode	3rd-Order PIM	5th-Order PIM	7th-Order PIM
GSM900	916–973○	897–992●	878–1011●
DCS1800	1780–1855○	1755–1880○	1730–1905●
DCS1800 (expanded)	1828–1894○	1806–1916○	1784–1938○
TD-F	1840–1960●	1800–2000●	1760–2040●
TD-A	1995–2040●	1980–2055●	1965–2070●
TD-E	2270–2420●	2220–2470●	2170–2520●
TD-D	2520–2670●	2470–2720●	2420–2770●
WLAN	2316.5–2567●	2233–2650.5●	2149.5–2734●

Note: ○ indicates that it does not fall into the system receiving band; ● indicates that it falls into the system receiving band.

1.2.2 Single System PIM Interference Scenario

1. Analysis of PIM Interference in China Mobile's Single System

   Currently, China Mobile's main systems include CSM900, DSC1800, TD-SCDMA (F/A/E/D), TD-LTE (F/A/E/D) and WLAN. For these single systems, combined with the spectrum usage listed in Table 1.1, Table 1.2 lists the single system PIM product of the system operated by China Mobile.

   As can be seen from Table 1.2, for Frequency Division Duplexing (FDD) systems (i.e. GSM900 and DSC1800 systems), the 3rd-order PIM products at the transmitting frequency point will not fall into the receiving band of this system, but the 5th- and 7th-order PIM products will fall into its system. For Time Division Duplexing (TDD) systems (such as TD-SCDMA and TD-LTE systems working in the F/A/E/D band), the 3rd-, 5th- and 7th-order PIM products at the transmitting frequency point will not fall into the receiving band of this system. For Time Division Duplexing (TDD) system (such as TD-SCDMA and TD-LTE system working in F/A/E/D band), the 3rd-, 5th- and 7th-order PIM products at the transmitting frequency will fall into the system band, but no effective interference will be generated since the TDD system operates in transmitting and receiving time division mode.

2. China Unicom Single System PIM Interference Analysis

   At present, the main wireless communication systems operated by China Unicom are GSM900, DSC1800, WCDMA, FDD-LTE and WLAN. For these single systems, the analysis combined with the spectrum usage listed in Table 1.1 shows that the low-order (3rd-, 5th- and 7th-order) PIM products in the downlink transmit band of GSM900, DSC1800 and WCDMA systems operated by China Unicom will not fall into their uplink receiving band, and the low-order (3rd-, 5th- and 7th-order) PIM products of WLAN systems will not interfere with this system.

3. China Telecom Single System PIM Interference Analysis

   At present, the main wireless communication systems operated by China Telecom are CDMA800, CDMA2000, FDD-LTE and WLAN. For these single systems, combined with the spectrum usage listed in Table 1.1, the analysis shows that the low-order (3rd-, 5th- and 7th-order) PIM products of CDMA800 and CDMA2000 downlink transmit bands will not fall into their uplink receiving bands, and the PIM products of WLAN system will not affect the operation of this system.

1.2.3 PIM Analysis of Microwave Components

In modern mobile communication networks, the signals of several channels (each with a power between a few and tens of watts) are generally transmitted through a pair of transmitting antennas, which are used simultaneously as receiving antennas (transceiver common status) or at least located in one of the receiving antenna attachments, so that the receiving channel must be ensured to be unaffected by the transmitting channel. On the surface, this seems unlikely to be a problem, because the respective channels are strictly isolated from each other. However, because the passive components must have nonlinear resistance, so, for a certain period, the current of several carrier channels flowing through such a resistor at the same time may quickly form interference in the form of mixed products, and then the interference signal reaches the receiving channel directly or indirectly or reaches the receiving antenna through the transmitting antenna. For example, the general input amplifier sensitivity is up to 0.01 pW/−110 dBm, and an interference signal 153 dB lower than the 20 W/43 dBm power transmitter (a difference of 15 orders of magnitude) is enough to cause one or more receiving channels to fail. With the increasing number of mobile communication networks and mobile communication base stations, various frequency components in space are becoming increasingly complex. They are superimposed and mixed with each other, and when certain conditions are met between several communication networks in the same area, it will cause interference to the communication system. With the increasing amount of voice and data information within the limited bandwidth through the mobile communication system, PIM interference has become an important factor limiting the capacity of the system.

In a typical ground mobile communication system, PIM sources are widely distributed, as shown in Figure 1.4. There are two main types of causes of PIM nonlinearity – contact nonlinearity (loose, oxidized, contaminated metal connection joints) and material nonlinearity (bulk materials, such as ferromagnetic components, carbon fibers), which exhibit nonlinear current-voltage characteristics.

In the long-term service of base stations, environmental pollution, corrosion, oxidation, etc. can easily cause PIM of passive components, so PIM testing of base stations should be carried out regularly; Figure 1.5 shows a typical urban base station.

In the context of increasing public awareness of environmental protection, coconstruction and sharing of base station facilities (including indoor distribution systems) have become an overall trend in the domestic and international communication industry. However, while reducing the costs, coconstruction and sharing have also made the problem of interference between them increasingly prominent.

The analysis of the 3rd-order PIM index of passive components in China started relatively late. For many years, the requirement of industry for third-order PIM suppression capability in passive components is −120 dBc@2×43 dBm, but with the increase of network load, this index can no longer meet the needs of network performance. In 2014, China

FIGURE 1.5   Typical base stations in cities.

Mobile randomly tested 2057 antennas and passive components in the east, middle and west of China; the pass rate of smart antennas is about 91%, while that of dual-frequency ESC dual-polarized antennas is the lowest, only about 30%. The main reason for the low pass rate is the low pass rate of PIM inhibition index. With the rapid development of communication technology, especially the increase of communication frequency of 5G antenna, and the increase of voice and data signal capacity, the factors that have less impact on signal before are also paid more and more attention, PIM is one of them. China Mobile has increased its requirements for PIM, and has issued standards such as "China Mobile Communications Enterprise Standards" and made special requirements for PIM; the EU's mobile communication system-related standards also include PIM as an index that must be assessed.

1.3 PROGRESS OF RESEARCH ON PIM OF SPACECRAFT MICROWAVE COMPONENTS

For satellite communication system, PIM is one of the key issues related to the success or failure of the system. When the PIM level is low, it will raise the bottom noise of the received signal, so that the signal-to-noise ratio of the receiver is reduced and the bit error rate is increased; when the PIM level increases further, it will affect the normal operation of the entire communication system, forcing the use of reduced power or that of subchannel; in severe cases, the PIM products will flood the received signal, resulting in channel blockage and communication interruption, and the entire system is paralyzed. Among the microwave components of the spacecraft payload, the microwave components that can usually

FIGURE 1.6   Diplexer.

FIGURE 1.7   Cable joint.

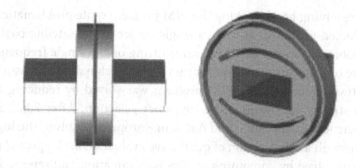

FIGURE 1.8 Waveguide flange.

generate PIM include transceiver duplexers (Figure 1.6), cable connectors (Figure 1.7), waveguide flange (Figure 1.8) and various antenna feeds and reflective surfaces.

PIM exists in any transceiver-shared system and exists in both ground equipment and space equipment. Because of the far distance and attenuation of space equipment and since the ratio of transmit power to receiver sensitivity is higher than that required for terrestrial equipment, the requirements for PIM are higher in satellite communication systems, and the consequences are serious particularly once the impact on satellite communication is produced. In a typical terrestrial WCDMA communication system, the transmit power is generally 43 dBm, the receiver requirement for PIM level is about −110 dBm, and the difference between transmit power and PIM level is 153 dB; however, in the space communication system, due to the long distance between the satellite and the ground and the large attenuation, the transmission power on the satellite is increased, and the receiving sensitivity is improved. Therefore, the difference between the transmission power and the PIM level is usually higher than that on the ground. The existence of extremely weak nonlinearity can also cause the PIM interference signal to flooding the received signal. Individual high-power terrestrial systems with PIM problems can be solved by using two antennas for transmitting and receiving and by using the method of space isolation, but it is difficult to use terrestrial measures to avoid this problem on satellites due to the limitation of space platform resources.

PIM has become a key concern in satellite development. The United States launched five mobile communication satellites in nearly 10 years after 1975, and the first four were severely affected by PIM. The 3rd-order PIM product of the U.S. fleet communication satellite FLTSATCOM – when the whole satellite entered the flight model debugging, there was a problem of the 3rd-order PIM product falling into the receiving band and was forced to switch to a separate transceiver program, which led to a 36-month delay in the launch of the whole satellite. The U.S. communications satellite MILSTAR-II adopted a separate transceiver scheme for the UHF band in order to avoid the 13th-order PIM products from falling into the receiving band. The 43rd-order PIM products of MARECS and the 27th-order PIM products of the international communication satellite INTELSAT all interfere with the satellite receiving signal, causing the satellite to reduce power usage and even scrap. The OPTUS C1 satellite launched in 2003 did not have a prior PIM constraint and risk assessment for the entire satellite, resulting in both 3rd and 23rd-order PIM products falling into the receiving band, making the PIM problem quite problematic. Experiments have been conducted to show that when a single device in the satellite payload is operating, the PIM problem does not occur by transmitting only a single frequency signal, and PIM interference only occurs when two (or more) carrier signals are transmitted in a multicamera joint test. Eventually, the PIM problem was solved by reducing the subcarrier bandwidth and strictly constraining the frequency selection. After 2005, Space Systems Loral, Orbital Sciences Corporation and Astrium Europe have solved the high-power load PIM problem through a combination of qualitative evaluation and repeated tests. The PIM problem has been solved by combining qualitative evaluation and repeated experiments, and some satellites have adopted the transceiver mode, which has significantly improved the effectiveness and represents the development direction of satellite payload technology.

China space technology has entered a new stage; one of the signs is the expansion of the waveband and the increase of power. As the saturation power per channel has increased from a few watts or a dozen watts before to hundreds of watts now, the PIM problem has become increasingly serious, so there is an urgent need to study how to effectively suppress the PIM effect in the satellite system to ensure that the satellite system can work properly during its life cycle.

With the rapid development of a new generation of mobile communication satellites, PIM problems are inevitable. Figure 1.9 shows the schematic diagram of satellite-ground communication, and the PIM products enter the receiver with the received signal, which cannot be solved by the traditional filtering and isolation methods. In order to meet the requirements of the PIM, the current research mainly focuses on the optimization of the overall system and its components from the perspective of the generation mechanism, including reasonable selection of the transceiver band, trying to avoid the low-order PIM products fall into the uplink band, avoiding the use of ferrite or ferromagnetic materials with strong nonlinear characteristics and making monolithic hardware without oxide layers or contamination-free films on metal surfaces or inside metal sheets. Trim the "metal-nonmetal-metal" contact surface to enhance the conductivity of the contact surface and reduce its nonlinear effect.

Since the PIM problem is of great importance to satellites, the international research on PIM is extensive and in-depth. With the development of China's space technology, the

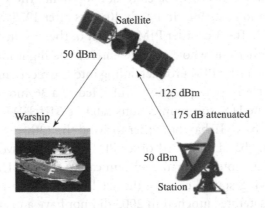

FIGURE 1.9 Schematic diagram of satellite-to-ground communication.

interference problems caused by PIM are becoming more and more prominent as the frequency bands used by satellites are expanding and the on-satellite power is increasing. Therefore, it is necessary to carry out theoretical research and engineering practice of PIM to protect the future development of multicarrier satellite systems.

1.4 SUMMARY

The PIM problem was first discovered in long-range naval communication, and then appeared one after another in satellite communication and ground base station communication. From the discovery of PIM, understanding PIM, and studying the mechanism of PIM generation, to the analysis of PIM nonlinearity using mathematical tools, and then to the study of efficient PIM measurement and PIM suppression techniques, it has taken more than half a century, and a lot of work has been carried out to study the generation mechanism, analysis and calculation methods, measurement techniques and suppression techniques. From the search of papers and patents (Figures 1.10 and 1.11), it can be seen that PIM problem has experienced three periods: budding period, rapid development period and stable development period. From the timeline, we can see that the rapid development period is accompanied by the rapid development of satellite technology and mobile communication technology. In the context of the development trend of "space-ground-object" integrated communication and the increasing spectrum resource constraint, the PIM problem will become more and more prominent, and microwave components and system-level PIM will face new challenges. This book will provide a comprehensive introduction to the mechanism, analysis and evaluation, detection and localization and suppression techniques for microwave components of spacecraft PIM, which will be an important reference for ground mobile communication systems.

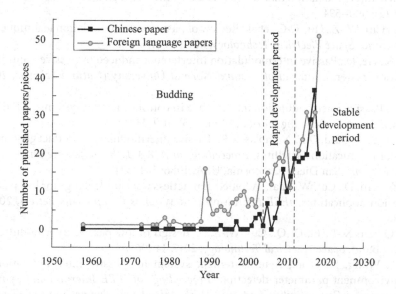

FIGURE 1.10 Statistics on the number of papers published in China and abroad. (Retrieved in August 2019.)

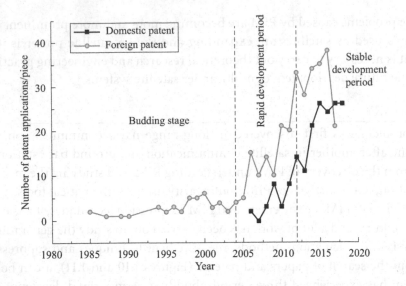

FIGURE 1.11 Statistics on the number of patent applications in China and abroad. (Retrieved in August 2019.)

BIBLIOGRAPHY

1. Zhang, L., Wang, H., He, S., et al., A segmented polynomial model to evaluate passive inter-modulation products from low - order pim measurements. *IEEE Microwave and Wireless Components Letters*, 2019, 29 (1): 14–16.
2. Zhang, L., Li, Y., Lin, S., et al., Numerical simulation and analysis of passive intermodulation caused by multipaction. *Physics of Plasmas*, 2018, 25 (8): 082301 (1)–082301 (7).
3. Zhang, L., Wang, H., et al., A composite exponential model to characterize nonlinearity caus-ing passive intermodulation interference. *IEEE Transactions on Electromagnetic Compatibility*, 2019, 61 (2): 590–594.
4. Li, X.-X., Cui, W.-Z., Hu, T.-C., et al., Review of passive intermodulation techniquesand devel-opment trend. *Space Electrical Technology*, 2017, 14 (4): 1–6.
5. Zhang, S., Ge, D., Passive intermodulation interference induced by passive nonlinear in com-munication systems. *Journal of Shaanxi Normal University (Natural Science)*, 2004, 32 (1): 58–62.
6. Lui, P.L., Passive intermodulation interference in communication systems. *IEE Electronics & Communication Engineering Journal*, 1990, 2 (3): 109–118.
7. Hoeber, C.F., Pollard, D.L., Nicholas, R.R., Passive intermodulation product generation in high power communications satellites. *Proceedings of AIAA 11th Conference on Communication Satellite Systems*. San Diego, California, USA, 1986: 361–374.
8. Chen, X., Sun, D., Cui, W., et al., A folded contactless waveguide flange for low passive - inter-modulation applications. *IEEE Microwave and Wireless Components Letters*, 2018, 28 (10): 864–866.
9. Wang, Q., Di, X.-F., Li, Q.-Q., Cui, W.-Z., A design of low-passive intermodulation coaxial filter in S-band. *Space Electrical Technology*, 2017, 14 (06):49–53.
10. Cui, W., Wang, R., et al., A measurement system for passive intermodulation with real time environment parameter detection. *Proceedings of IEEE International Symposium On Electromagnetic Compatibility And 2018 IEEE Asia-Pacific Symposium On Electromagnetic Compatibility (Emc / Apemc)*. Davos, Switzerland, 2018: 956–958.

11. Chen, X., He, Y., Cui, W., Broadband dual - port intermodulation generator for passive inter-modulation measurements. *IEEE Microwave and Wireless Components Letters*, 2017, 27 (5): 518–520.

12. Chen, X., Shuang, L., Sun, D., et al., Suspended contactless low passive intermodulation tran-sition of waveguide flange. *Journal of Xi'An Jiaotong University*, 2020, 54 (5): 117–123.

13. Bolli, P., Selleri, S., Pelosi, G., Passive intermodulation on large reflector antennas. *IEEE Antenna's And Propagation Magazine*, 2002, 44 (5): 13–20.

14. Chen, X., Cui, W., Time-domain detection of passive intermodulation and its application in PIM localization of mesh reflector antennas. *China Academic Journal Electronic Publishing House. Proceedings of Microwave and Millimeter Conference*. 2018: 131–134.

15. Hienonen, S., Vainikainen, P., Raisanen, A.V., Sensitivity measurements of a passive inter-modulation near - field scanner. *IEEE Antenna's and Propagation Magazine*, 2003, 45 (4): 124–129.

16. Zhang, S.-Q., Study of Passive Intermodulation Interference at Microwave and RF Frequencies. Xidian University, 2004.

17. Boyhan, J.W., Lenzinc, H.F., Koduru, C., Satellite passive intermodulation: Systems consider-ations. *IEEE Transactions on Aerospace and Electronic Systems*, 1996, 32 (3): 1058–1064.

18. Wang, H., Liang, J., Wang, J., Zhang, C., Review of passive intermodulation in HPM condi-tion. *Journal of Microwaves*, 2005, 21: 1–6.

19. Henrie, J.J., A Study of Passive Intermodulation in Coaxial Cable Connectors. West Lafayette: Purdue University, 2009.

20. Enc, K.Y., Stern, T.E., The order - and - type prediction problem arising from passive inter-modulation interference in communications satellites. *IEEE Transactions on Communications*, 1981, 29 (5): 549–555.

21. Ye, M., He, Y., Wang, X., Cui, W., Study of passive intermodulation of microwave waveguide connection based on rough surface model. *Journal of Microwaves*, 2010, 26(4): 65–69.

22. Khattab, T.E., The effect of high power at microwave frequencies on the linearity of non - polar dielectrics in space RF component. *International Journal of Communications Network & System Sciences*, 2015, 8 (2): 11–18.

23. Bayrak, M., Benson, F.A., Intermodulation products from nonlinearities in transmission lines and connectors at microwave frequencies. *Proceedings of the Institution of Electrical Engineers*, 1975, 122 (4): 361–367.

24. Ye, M., Wu, C., He, Y., et al., Experimental research on passive intermodulation character-istics of s-band waveguide to coaxial adapter. *Chinese Journal of Radio Science*, 2015, 30 (1): 183–187.

25. Wilkerson, J.R., Kilcore, I.M., Card, K.G., et al., Passive intermodulation distortion in anten-nas. *IEEE Transactions on Antennas & Propagation*, 2015, 63 (2): 474–482.

26. Jiang, J., Li, T., Mei, J., Wang, H., Passive intermodulation analysis of coupled electro-thermal microwave loads. *Journal of Xidian University (Natural Science)*, 2016, 43(3): 179–184.

27. Zhao, P., Study of Passive Intermodulation Interference in Uhf Band of Wireless Communication Systems. Beijing University of Post and Telecommunications.

The Generation Mechanism of Passive Intermodulation of Microwave Components

Chunjiang Bai, Yun He, and Yongning He

CONTENTS

DOI: 10.1201/9781003269953-2

2.1 GENERAL OVERVIEW

Generally, PIM exists in any system that is shared by sending and receiving, and it exists in both ground equipment and space equipment. In the transmitting system, although the amplitude of the PIM product is much lower than that of the transmitted signal, it will not affect the quality of the transmitted signal, but if these weak PIM products enter a highly sensitive receiver, it is very likely to exceed the thermal noise subband of the receiver, which will affect the normal operation of the system. For ground mobile communication systems, the transmission power is relatively low, and the receiver's requirements for PIM levels are also low; but in space communication systems, the distance between the satellite and the ground is large and the attenuation is large, which requires the transmission power on the satellite to increase, the receiving sensitivity is improved, so the receiver has higher requirements for PIM level, and even the extremely weak nonlinearity in the microwave components can cause the PIM interference signal to flood the received signal.

There are two types of PIM in the system: material nonlinearity and contact nonlinearity. Material nonlinearity refers to the inherent nonlinear characteristics of the material medium with nonlinear conductive characteristics, such as ferromagnetic materials; contact nonlinearity refers to the nonlinearity caused by contact with nonlinear current/voltage characteristics, such as metal contact with oxidation and corrosion surface. In order to avoid PIM, materials with weak nonlinearity are usually selected for component design. Contact nonlinearity is inevitable in microwave passive components and is the main source of electromagnetic interference. Therefore, this chapter will introduce the contact nonlinearity of microwave components and its relationship with PIM.

In view of the fact that most of the materials used in the microwave components of the space communication system are aluminum alloy silver-plated, this chapter will take the Ag-Ag contacts that are abundant in the space microwave components as the research object and combine the main carrier nonlinear transport processes, such as quantum tunneling and hot electron emission, to determine the main nonlinear transport physical mechanisms of metal contact junctions, so as to obtain the nonlinear current density equation for microasperity contacts; by combining the nonlinear current characteristic equation of single point structure and microasperity contact Monte Carlo method of microwave part contact junction, the statistical analysis of nonlinear surface current of contact interface is realized, and the influence law of surface material, surface morphology, and connection pressure on the nonlinear characteristics of surface contact are systematically studied. In addition, this chapter will introduce the electrothermal coupling effect and analyze the PIM generation mechanism based on the electrothermal coupling effect, explore and initially reveal the complex formation mechanism of PIM and provide scientific theoretical basis for the PIM analysis and the evaluation and the suppression of microwave components.

2.2 STUDY OF THE PHYSICAL MECHANISM OF NONLINEAR TRANSPORT IN SINGLE-POINT STRUCTURES

2.2.1 Experimental Study of the Surface Composition of Aluminum Alloy Silver-Plated Specimens

In order to determine the type of oxides and stains on the metal surface of the actual microwave components, this chapter selects aluminum alloy silver-plated specimens with a similar microwave component process to compare the changes in the surface XPS (X-ray photoelectron spectroscopy) of each specimen before and after cleaning under ultra-high vacuum conditions. The surface compositions obtained from the tests are shown in Table 2.1.

Under the condition that the specimen was not cleaned with organic solvent, the XPS analysis of the surface of the silver-plated specimen showed that there was organic molecular staining on the surface; after the organic solvent cleaning, the proportion of C component on the surface of the silver-plated specimen decreased, and the proportion of Ag and O components increased; after the continued Ar plasma cleaning, the C component almost completely disappeared, but the proportion of Ag and O components was close to the atomic ratio of Ag_2O. It can be seen that the surface of the aluminum alloy silver-plated specimen is stained with carbon, hydrogen and oxygen molecules, and an Ag_2O layer is formed on the surface of the silver-plated layer.

Then, Scanning Electron Microscopy (SEM) observations were made on the surface of the discolored silver-plated specimen placed in an ordinary environment, as shown in Figure 2.1. Compared with the original smoother surface, a layer of polycrystalline Ag_2O particulate matter has appeared on the surface of the discolored specimen, and the oxide grain size is about 20nm. Therefore, the processing, placement and assembly of silver-plated microwave components should be carried out in a dry and noncorrosive environment, and a humid environment should be avoided as much as possible, which is beneficial to prevent discoloration and corrosion.

TABLE 2.1 Surface Composition of Aluminum Alloy Silver-Plated Specimens Before and After Cleaning (XPS) %

Surface Elements	Ag	C	O	Cr
Aluminum alloy silver-plated specimens	13.7	34.4	43.4	8.5
Organic solvent cleaning	18.3	27.5	42.4	11.9
Organic solvent cleaning plus Ar plasma cleaning	79.3	0	20.7	0

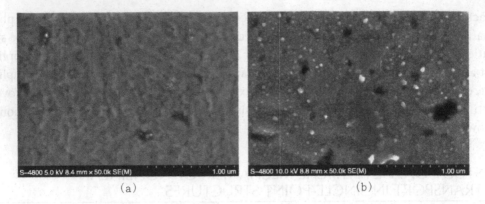

(a)   (b)

FIGURE 2.1   SEM images of the original surface of the silver-plated specimen and the surface of the placed oxidation discoloration. (a) Original surface. (b) Placed oxidation discoloration surface.

2.2.2 MOM Structure Interface Potential Barrier Model and Carrier Transport Mechanism

In the oxide films formed by the natural low-temperature oxidation process of metal plating of actual microwave components, there are certain concentrations of donor and acceptor defects. Next, the potential barrier model, model parameters, and the main transport equations for analyzing MOM (Metal-Oxide-Metal) structures based on ultrathin oxide layers on metal surfaces are discussed according to various cases based on the types of defects present in the oxide.

2.2.2.1 Structural Analysis of MOM with Predominant Donor Defects in the Oxide Layer

Here, the formation of Ag_2O on the silver surface is analyzed as an example. The defects generated during the formation of Ag_2O are Ag gaps, oxygen vacancies and impurity shallow donor, among which the shallow donor defects ionize into n-type semiconductors, and compared to intrinsic semiconductors, the Fermi energy level EF guides the band bottom closer and the metal work function is larger than the oxide work function, in which case the MO (metal oxide) interface forms a Schottky contact barrier and the tunneling current and Schottky thermal emission currents are the main current mechanisms. The thermal equilibrium barrier model of the weak n-type Ag_2O MOM structure and its barrier model under the bias voltage (referred to as "bias voltage") are presented here as shown in Figure 2.2.

For weak n-type oxides, it can be approximated as a dielectric layer with a uniform field in the oxide layer under the applied voltage; for strong n-type oxides, the oxide depletion zone can shield the electric field and will mainly land on the interface of the reverse bias under the condition of a small applied voltage, and the depletion layer is broadened toward the oxide, which can also be approximated as a uniform field condition under the thin oxide layer condition.

Under the condition that the image force effect is not considered, we have

$$\phi_{bn} = \phi_m - \chi_{ox} \tag{2.1}$$

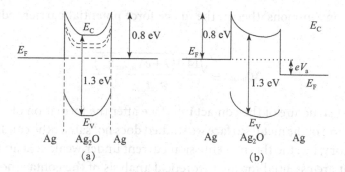

FIGURE 2.2 Potential barrier model of weak n-type Ag_2O MOM structure under thermal equilibrium and bias pressure conditions. (a) Without bias voltage. (b) With bias voltage. E_c – conduction band energy level; E_F – Fermi energy level; E_v – valence band energy level.

where Φ_{bn} is the height of the potential barrier considering the image force effect; Φ_m is the height of the potential barrier between the metal and the semiconductor contact; χ_{ox} is the electron affinity energy of the oxide.

According to the theory of hot electron emission in solid state electronics, the equation of hot electron emission current of MOM structure is obtained as

$$J_{Th}(V_a) = AT^2 \exp\left(-\frac{e\phi_{b0}}{k_BT}\right)\left(1 - \exp\left(-\frac{eV_a}{k_BT}\right)\right) \tag{2.2}$$

where V_a is bias voltage (V), A is an effective Richardson's constant, taking the value of $\frac{4\pi mek_B^2}{h^3}$, m is the electron mass, k_B is Boltzmann's constant, h is Planck's constant, T represents temperature (K) and Φ_{b0} is the height of the potential barrier (eV) of the MOM structure without considering the image effect.

Considering the image force effect, a more accurate equation for the hot electron emission current is obtained as

$$J_{Th,\,image}(V_a) = AT^2 \exp\left(-\frac{e\left(\phi_{b0} - \Delta\phi_{image}\right)}{k_BT}\right)\left(1 - \exp\left(-\frac{eV_a}{k_BT}\right)\right) \tag{2.3}$$

Simmons assumes that the space charge confinement effect can be neglected, the field strength between the oxide contact electrodes is large enough and the electric field distribution on the oxide between the two electrodes is a uniform field condition under the applied bias condition, and the classical Simmons' hot electron emission current equation is obtained. According to Simmons' theory, the hot electron emission equation for the MOM structure is

$$J_{Th,\,Simmons}(V_a) = AT^2 \exp\left(\frac{\sqrt{14.4(7 + eV_a\varepsilon_{ox}t_{ox})} - \varepsilon_r t_{ox}\phi_{b0}}{\varepsilon_{ox}t_{ox}k_BT}\right)\left(1 - \exp\left(-\frac{eV_a}{k_BT}\right)\right) \tag{2.4}$$

where ε_{ox} is the relative permittivity of the oxide layer, and t_{ox} is the thickness of the oxide layer.

That is, based on Simmons' theory, the image force potential barrier reduction effect is approximately

$$\Delta\phi_{image} = \frac{\sqrt{14.4\left(7 + eV_a\varepsilon_{ox}t_{ox}\right)}}{\varepsilon_{ox}t_{ox}} \qquad (2.5)$$

In fact, the MOM structure of the contact interface after the formation of ultrathin oxides by natural oxidation of the metal surface we studied does not satisfy the conditions assumed in Simmons' theory, i.e., the thermal emission current under weak field and the tunneling current behind it are essential for the theoretical analysis of the contact nonlinearity. The derivation of the reduction effect of the image force barrier in Simmons theory has been approximated and fitted. We use a more accurate model of the reduction effect of the image force barrier developed based on the Coulomb effect after Simmons' theory. As shown in Figure 2.3, the potential function in the oxide under uniform field conditions is

$$-\phi(x) = -\frac{e}{16\pi\varepsilon_{ox}\varepsilon_0 x} - Ex \qquad (2.6)$$

where x is the position in the oxide layer starting from metal oxide; $-\dfrac{e}{16\pi\varepsilon_{ox}\varepsilon_0 x}$ is the potential function derived from the Coulomb effect of the image image charge, ε_0 is the vacuum dielectric constant and E is the electric field strength,

$$E = \frac{V_a}{t_{ox}} \qquad (2.7)$$

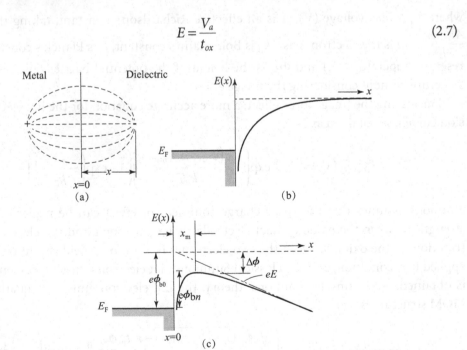

FIGURE 2.3 Metal-semiconductor contact potential barrier reduction principle. (a) Charge image principle schematic diagram. (b) Metal-semiconductor contact potential barrier model. (c) Considering the image force of metal-semiconductor contact potential barrier model.

According to $\dfrac{d(e\phi(x))}{dx} = 0$, the highest position of the potential barrier is obtained at x_m as

$$x_m = \sqrt{\dfrac{e}{16\pi\varepsilon_{ox}\varepsilon_0 E}} \tag{2.8}$$

Thus, the magnitude of the reduction of the potential barrier due to the image force is obtained as

$$\Delta\phi_{image} = \sqrt{\dfrac{eE}{4\pi\varepsilon_{ox}\varepsilon_0}} \tag{2.9}$$

Substituting this into Equation (2.3), the equation for the hot electron emission current considering the image force effect is

$$J_{Th,ox}(V_a) = AT^2 \exp\left(-\dfrac{e\phi_{b0}}{k_B T}\right)\exp\left(\sqrt{\dfrac{e^3 V_a}{4\pi\varepsilon_{ox}\varepsilon_0 t_{ox}}}/(k_B T)\right)\left(1-\exp\left(-\dfrac{eV_a}{k_B T}\right)\right) \tag{2.10}$$

Taking the parameters of Ag$_2$O as an example, with $t_{ox}= 10\,\text{nm}$, $V_a= 100\,\text{mV}$, $\varepsilon_{ox}=3.2$, we have $x_m=3.35\,\text{nm}$.

$$\Delta\phi = \sqrt{\dfrac{eE}{4\pi\varepsilon_{ox}\varepsilon_0}} = 0.067\,\text{V}$$

In fact, Equeation (2.9) holds only if $x_m \leq t_{ox}$ (i.e., if the oxide layer is thick enough), when the hot electron emission current equation (2.10) must satisfy the condition $x_m \leq t_{ox}$, i.e., it holds only under the applied bias voltage of $V_a \geq \dfrac{e}{16\pi\varepsilon_{ox}\varepsilon_0 t_{ox}}$. In fact, the naturally occurring oxide layer is so thin that when the condition $x_m = \sqrt{\dfrac{e}{16\pi\varepsilon_{ox}\varepsilon_0 E}} \geq t_{ox}$ is satisfied (i.e., under the field strength of $V_a \leq \dfrac{e}{16\pi\varepsilon_{ox}\varepsilon_0 t_{ox}}$), the potential barrier is reduced as

$$\Delta\phi = \dfrac{e}{16\pi\varepsilon_{0x}\varepsilon_0 t_{ox}} + V_a \tag{2.11}$$

Substituting this into equation (2.10), we have

$$J_{Th,ox}(V_a) = AT^2 \exp\left(-\dfrac{e\phi_{b0}}{k_B T}\right)\exp\left(\dfrac{e^2}{16\pi\varepsilon_{ox}\varepsilon_0 t_{ox}} + eV_a\right)\left(1-\exp\left(-\dfrac{eV_a}{k_B T}\right)\right)$$

$$= AT^2 \exp\left(-\dfrac{e}{k_B T}\left(\phi_{b0} - \dfrac{e}{16\pi\varepsilon_{ox}\varepsilon_0 t_{ox}}\right)\right)\left(\exp\left(\dfrac{eV_a}{k_B T}\right)-1\right) \tag{2.12}$$

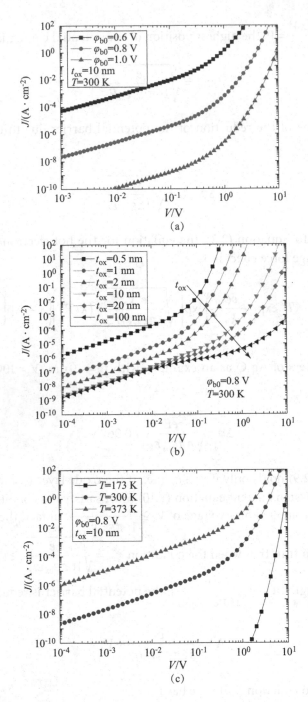

FIGURE 2.4 Thermal emission current density *J-V* characteristics. (a) Effect of different contact potentials on current density. (b) Effect of different oxide layer thickness on current density. (c) Effect of different temperatures on current density.

In contrast to the strong dependence of the thermal emission current process (Figure 2.4) on the oxide barrier height and the applied bias voltage, the electron tunneling process of the thin oxide barrier of the MOM structure calculated according to the quantum tunneling theory not only depends on the oxide barrier height but also has a strong dependence on the barrier shape (i.e., the barrier as a function of the tunneling direction). According to Simmons' theory based on the WKB[J] approximation, a general expression for the tunneling current under the applied bias voltage V_a can be derived as

$$J_{T_u}(V_a) = J_0 \bar{\phi} \exp\left(-K\bar{\phi}^{1/2}\right) - J_0\left(\bar{\phi} + eV_a\right)\exp\left(-K\left(\bar{\phi} + eV_a\right)^{1/2}\right) \tag{2.13}$$

where $\bar{\phi}$ is the average value of the height of the potential barrier above the Fermi energy level of the left electrode with negative pressure.

$$J_0 = \frac{e}{2\pi h t_{ox}^2} \tag{2.14}$$

$$\bar{\phi} = \frac{1}{t_{ox}} \int_{s_1}^{s_2} \phi(x)dx \tag{2.15}$$

where s_1, s_2 are the positions of the two sides of the oxide layer.

$$K = \left(4\pi t_{ox}/h\right)\sqrt{2m} \tag{2.16}$$

As shown in Figure 2.5, the first term in Equation (2.13) can be regarded as the current flowing from the left electrode to the right electrode, and the second term is the current flowing from the right electrode to the left electrode, and the two are subtracted to obtain the net quantum tunneling current flowing from the left electrode to the right electrode under the bias voltage V_a.

Considering the image force and the condition that the field distribution on the oxide layer is homogeneous, Simmons theory gives the following formula for calculating the quantum tunneling current

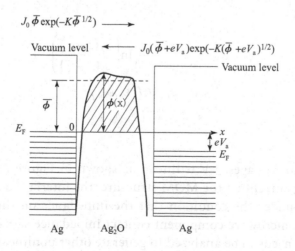

FIGURE 2.5   Variation of potential vs. current under bias voltage.

$$J_{TU}(V_a) = \frac{6.2 \times 10^{10}}{t_{ox}^2} \left(\phi_I \exp\left(-1.025 t_{ox} \phi_I^{1/2}\right) \right.$$

$$\left. -(\phi_I + V_a) \exp\left(-1.025 t_{ox}(\phi_I + V_a)^{1/2}\right) \right) \tag{2.17}$$

where

$$\phi_1 = \phi_{b0} - \frac{V_a}{2s}(s_1 + s_2) - \frac{5.75}{\varepsilon_{ox}(s_2 - s_1)} \ln\left(\frac{s_2(s - s_1)}{s_1(s - s_2)}\right)$$

$$\text{s.t.} \quad s_1 = 6/(\varepsilon_{ox}\phi_{b0}) \tag{2.18}$$

$$s_2 = \begin{cases} s\left(1 - 46/(3\phi_{b0}\varepsilon_{ox}s + 20 - 2V_a\varepsilon_{ox}s)\right) + 6/(\varepsilon_{ox}\phi_{b0}), & V_a < \phi_{b0} \\ (\phi_{b0}\varepsilon_{ox}s - 28)/(\varepsilon_{ox}V_a), & V_a > \phi_{b0} \end{cases}$$

where s is the position in the oxide layer starting from the metal oxide layer interface.

In addition, since the RF connection contact nonlinear current is an extremely weak signal, it is also necessary to consider the $J-V$ relationship at very small voltages, i.e., under the condition that $V_a \approx 0$, we have

$$J_{TU}(V_a) = \frac{3.16 \times 10^{10}}{t_{ox}^2} \left(\phi_I \exp\left(-1.025 t_{ox} \phi_L^{1/2}\right) \right.$$

$$\left. -(\phi_I + V_a) \exp\left(-1.025 t_{ox}(\phi_I + V_a)^{1/2}\right) \right) \tag{2.19}$$

where

$$\phi_1 = \phi_{b0} - \frac{5.75}{\varepsilon_{ox}(s_2 - s_1)} \ln\left(\frac{s_2(s - s_1)}{s_1(s - s_2)}\right)$$

$$\text{s.t.} \quad s_1 = 6/(\varepsilon_{ox}\phi_{b0}) \tag{2.20}$$

$$s_2 = 1 - 6/(\varepsilon_{ox}\phi_{b0})$$

In fact, at very small voltages, J-V is linear, as shown in Figure 2.6. That is, even if there is a tunneling current on the MOM structure, the interface oxide still resembles a linear resistance under the condition that the impedance on the contact junction is small, and if the microwave component contact impedance satisfies this condition, then this condition needs to be analyzed to generate other nonlinear characteristics of the PIM.

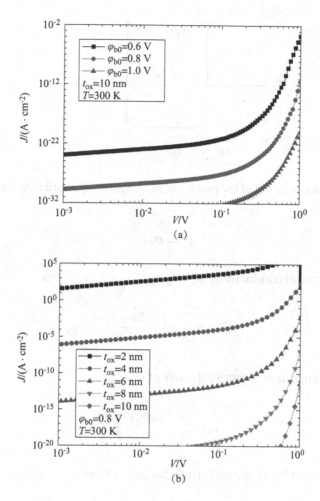

FIGURE 2.6 *J-V* characteristic curve of tunneling current. (a) Effect of different barrier heights on current density. (b) Effect of different oxide layer thickness on current density.

2.2.2.2 Potential Model of MOM Structure with Predominant Host-Type Defects in the Oxide Layer and its Transport Equation

The host-acceptor type defects in Ag_2O are mainly silver vacancies formed during the preparation of silver oxide and the introduction of host-acceptor impurities. When a large number of host-acceptor type defects are present in Ag_2O, it is called p-type semiconductor. Under the weak p-type condition, the metal work function φ_m changes from greater than the oxide work function φ_{ox} to equal to the oxide work function as the hole concentration increases. According to the semiconductor theory, under the condition of $\varphi_m > \varphi_{ox}$, a p-type antiblocking layer is formed at the metal-oxide interface, as shown in Figure 2.7.

The antibarrier layer is actually a very thin and high conductivity layer, which has little effect on the contact resistance of the oxide and the metal. That is, under the condition of applied bias, the conductivity process on the oxide layer is no longer influenced by the electrode contact potential barrier but is controlled by the bulk process. The dependence of its current density $JL0$ on the electric field strength E is

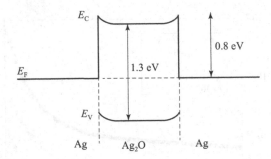

FIGURE 2.7   Schematic diagram of the process of forming a p-type antiblocking layer at the metal-oxide interface.

$$J_{L0} = \sigma_0 E \tag{2.21}$$

where σ_0 is the electrical conductivity, $\sigma_0 = p_{ox} e_u$,

$$p_{ox} = N_V \exp\left(-\frac{E_F - E_V}{k_B T}\right) \tag{2.22}$$

where N_v is the valence band effective density of states,

$$N_V = 2\left(\frac{2\pi m k_B T}{h^2}\right)^{3/2} \tag{2.23}$$

If $N_v = 2.5 \times 1019$ /cm^3 at RT is approximately taken, we have

$$p_{ox} = 2.5 \times 10^{19} \exp\left(-\frac{0.5}{0.0256}\right)$$

$$= 8.23 \times 10^{10} \text{ / cm}^3 \tag{2.24}$$

We take $u = 100$ cm^2/(V·s), the current density at RT can be obtained as

$$J_{L0} = \sigma_0 E = 8.23 \times 10^{10} \times 10^6 \times 1.6 \times 10^{-19} \times 100 \times 10^{-4} \frac{V_a}{t_{ox}}$$

$$= 1.3 \times 10^{-4} \frac{V_a}{t_{ox}} \left(A / m^2\right) \tag{2.25}$$

That is, J-V is linearly dependent on RT. However, the conductivity is a function of temperature, and the higher the temperature, the higher the concentration of intrinsic carriers, while under the condition that only lattice phonon scattering is considered, the higher the temperature, the lower the mobility. The nonlinear dependence of conductivity on temperature may lead to electrothermal coupling effect.

With the increase of the hole concentration in the silver oxide layer, the Fermi energy level approaches to the top of the valence band, and the dependence of the Fermi energy level on the hole concentration is as follows

$$E_{Fp} = E_V + k_B T \ln\left(\frac{N_V}{p_0}\right) \tag{2.26}$$

where p_0 is the hole density.

Under strong ionization conditions, $p_0 \approx NA$

$$E_{Fp} = E_V + k_B T \ln\left(\frac{N_V}{N_A}\right) \tag{2.27}$$

where NA is the acceptor concentration in Ag$_2$O.

The work function φ_{ox} of the oxide is

$$\phi_{ox} = \chi + E_g - k_B T \ln\left(\frac{N_V}{p_0}\right)$$

$$= 35 + 13 - k_B T \ln\left(\frac{N_V}{p_0}\right) \tag{2.28}$$

where X is the electron affinity of Ag$_2$O, and E_g is the forbidden band width.

When the hole concentration in the oxide is greater than $8.23 \times 1010/cm^3$, the work function of the metal is smaller than the oxide work function, and according to the semiconductor theory, the MO interface of the p-type oxide and the metal forms a barrier layer. The barrier layer formed at the MO interface of p-type silver oxide and metal at $\varphi_m = \varphi_{ox} = 4.5\,eV$ is given below. The potential barrier model of MOM structure under the condition of thermal equilibrium and under bias pressure is shown in Figure 2.8.

The height of the hole barrier φ_{bp} is determined by the following formula:

$$\phi_{bp} = \chi + E_g - \phi_m = 0.5\,eV \tag{2.29}$$

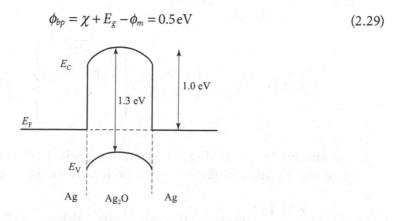

FIGURE 2.8   MOM structure barrier model of p-type silver oxide.

The natural oxidation process of the silver-plated layer of microwave components includes the formation of silver vacancies during the oxidation process. According to the low-temperature oxidation kinetics, it is preliminarily judged that there is a distribution of silver vacancies from metal-oxide layer to oxide layer-atmosphere in the Ag_2O formed by low-temperature thermal oxidation. It should be determined based on experimental research on the controllable low-temperature oxidation process.

2.2.3 Basic Nonlinear Equation of Metal Contact MOM Single Point Structure on Silver-Plated Surface

From the previous analysis, it is clear that the nonlinear current mechanisms affected by the interface potential barrier mainly include thermal emission current, direct tunneling current and F–N (Fowler–Nordheim) field emission tunneling current, and the nonlinear current mechanisms affected by the bulk effect mainly include P–F (Pool–Frenkel) thermal emission current and space-charge-limited current. For the actual MOM structure, the main mechanisms affecting the MOM nonlinear currents vary due to the film thickness of the oxide layer, defects in the film and temperature. In fact, the tunneling current only works when the oxide layer thickness is thin and decreases rapidly as the oxide layer thickness increases. Therefore, the combined characteristics of the nonlinear currents need to be given separately for the actual MOM structure for different cases.

2.2.3.1 Basic Nonlinear Equations for the MOM Single-point Structure

1. J-V relationship of thermal emission current considering the effect of image force barrier reduction.

When $V_a \leq \dfrac{e}{16\pi\varepsilon_{ox}\varepsilon_0 t_{ox}}$, we have

$$J_{Th,0x}(V_a) = AT^2 \exp\left(-\frac{e}{k_B T}\left(\phi_{b0} - \frac{e}{16\pi\varepsilon_{ox}\varepsilon_0 t_{ox}}\right)\right)\left(\exp\left(\frac{eV_a}{k_B T}\right) - 1\right) \quad (2.30)$$

When $V_a \geq \dfrac{e}{16\pi\varepsilon_{ox}\varepsilon_0 t_{ox}}$, we have

$$J_{Th,0x}(V_a) = AT^2 \exp\left(-\frac{e\phi_{b0}}{k_B T}\right)\exp\left(\frac{\sqrt{\dfrac{e^3 V_a}{4\pi\varepsilon_{ox}\varepsilon_0 t_{ox}}}}{k_B T}\right)\left(1 - \exp\left(-\frac{eV_a}{k_B T}\right)\right) \quad (2.31)$$

2. The formula for calculating the quantum tunneling current considering the image force and the field distribution on the oxide layer for the case of uniform field is

$$J_{TU}(V_a) = \frac{6.2\times10^{10}}{t_{OX}^2}\left(\phi_I \exp\left(-1.025 t_{OX}\phi_1^{1/2}\right) - (\phi_1 + V_a)\exp\left(-1.025 t_{OX}(\phi_1 + V_a)^{1/2}\right)\right) \quad (2.32)$$

where

$$\phi_1 = \phi_{b0} - \frac{V_a}{2s}(s_1 + s_2) - \frac{5.75}{\varepsilon_{ox}(s_2 - s_1)} \ln \frac{s_2(s - s_1)}{s_1(s - s_2)}$$

$$\text{s. t.} \quad s_1 = 6/(\varepsilon_{ox}\phi_{b0}) \tag{2.33}$$

$$s_2 = \begin{cases} s(1 - 46/(3\phi_{b0}\varepsilon_{ox}s + 20 - 2V_a\varepsilon_{ox}s)) + 6/(\varepsilon_{ox}\phi_{b0}), V_a < \phi_{b0} \\ (\phi_{b0}\varepsilon_{ox}s - 28)/(\varepsilon_{ox}V_a), \quad V_a > \phi_{b0} \end{cases}$$

3. F-N field emission current J-V characteristics:

$$J_{FN} = \frac{e^3}{8\pi h} \frac{V_a^2}{\phi_{b0}t_{ox}^2} \exp\left(-\frac{4\sqrt{2m^*}}{3e} \frac{\phi_{b0}^{3/2} t_{ox}}{V_a}\right) \tag{2.34}$$

4. P-F thermal emission current J-V relationship.

Considering the presence of traps in the oxide at a concentration of %, the activation energy of the trap is E_t.

$$J_{PF} = J_{L0} \exp\left(\sqrt{\frac{e^3}{4\pi(k_B T)^2 \varepsilon_{ox}\varepsilon_0}} E^{1/2}\right) \tag{2.35}$$

where $J_{L0} = e\mu N_C \sqrt{\dfrac{N_d}{N_t}} \exp\left(-\dfrac{E_t + E_d}{2k_B T}\right) E, \mu$ is the electron mobility, N_c is the effective state density of the conduction band, N_d is the donor impurity concentration, and E_d is the energy in band structure.

5. Space charge limiting current.

It is considered that there is a trap with concentration of N_t in the oxide, and the activation energy of the trap is E_t.

When $V_a \leq \dfrac{eN_t t_{ox}^2}{2\varepsilon_0 \varepsilon_r}$, we have

$$J_{SP} = \frac{9\varepsilon_0 \varepsilon_{ox} \mu V_a^2}{8t_{ox}^3} \tag{2.36}$$

When $V_a \geq \dfrac{eN_t t_{ox}^2}{2\varepsilon_0 \varepsilon_r}$, we have

$$J_{SP} = \frac{9\varepsilon_0 \varepsilon_{ox} \mu}{8t_{ox}^3} \Theta V_a^2 \tag{2.37}$$

where $\Theta = \dfrac{N_C}{N_t} \exp\left(-\dfrac{E_t}{k_B T}\right)$.

2.2.3.2 Nonlinear Current Synthesis Characteristics of MOM Single Point Structure

Here, a computational analysis is performed as an example of a silver-plated surface contact MOM structure. When the silver oxide layer is n-type or strong p-type, a barrier layer potential is formed at the Ag/Ag_2O interface. When the oxide layer is thin, the nonlinear current is mainly determined by the tunneling current, hot electron emission current and F-N field emission current. Figures 2.9 and 2.10 give the *J-V* relationship curves when the oxide layer is n-type and strong p-type, respectively, and the oxide layer thickness is taken as 4 and 5 nm, respectively. It can be seen that when the oxide layer is thin, the current is mainly affected by the tunneling current; when the oxide layer thickness increases, the tunneling current decreases rapidly and the thermal emission plays a major role.

For the weak p-type silver oxide layer, Ag and Ag_2O form an ohmic interface contact, and the nonlinear current is mainly affected by the receptor effect, mainly including the

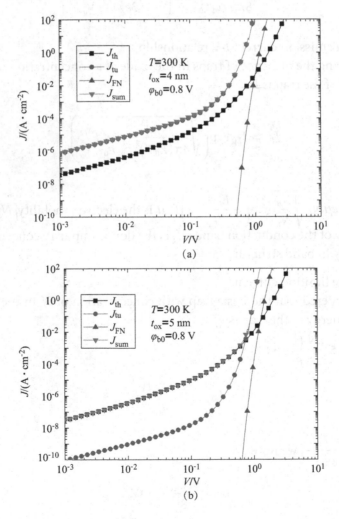

FIGURE 2.9 *J-V* characteristic curve of thin oxide layer with n-type barrier potential. (a) Current density for the oxide layer thickness of 4 nm. (b) Current density for the oxide layer thickness of 5 nm.

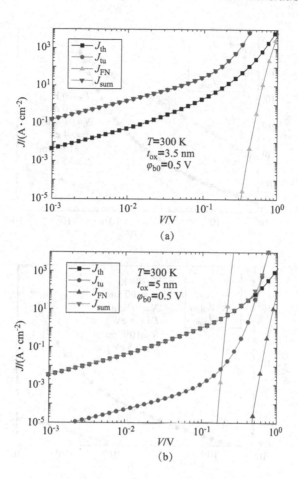

FIGURE 2.10 *J-V* characteristic curve of p-type barrier (strong p-Type) thin oxide layer. (a) Current density for the oxide layer thickness of 3.5 nm. (b) Current density for the oxide layer thickness of 5 nm.

P-F thermal emission current and the space charge limiting current. When the thickness of the oxide layer is thick and a barrier layer is formed between the metal and the oxide layer (strong p-type or n-type), the current mechanism of the contact interface (thermal emission current and F-N current) and body effect mechanism (P-F thermal emission current, space charge limiting current and linear current) are as shown in Figures 2.11 and 2.12.

2.3 MICROCONTACT MODELING OF MECHANICALLY CONNECTED JUNCTIONS OF MICROWAVE COMPONENTS

2.3.1 Classical Mechanical Model of Microwave Component Contact Based on Experimental Extraction of Surface Parameters

According to classical contact mechanics, three main parameters (microasperity surface density η, standard deviation of microasperity height σ and microasperity radius R) are used to describe the surface morphology of microwave components, which can be extracted from 3D morphology data from Laser Scanning Microscopy (LSM) or Atomic

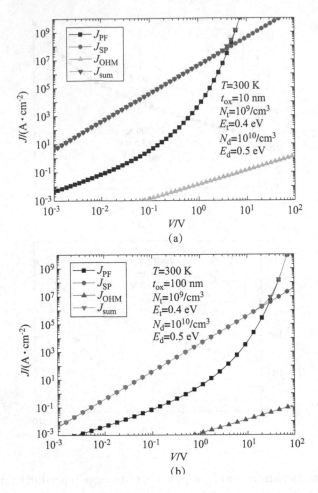

FIGURE 2.11 J-V Characteristic curve of ohmic interface contact (weak p-type). (a) Current density for the oxide layer thickness of 10 nm. (b) Current density for the oxide layer thickness of 100 nm.

Force Microscopy (AFM). In the classical model of contact mechanics, namely the GW (Greenwood Williamsons) model, the contact of two rough surfaces is equivalent to the contact of a smooth rigid plane with a rough surface with equivalent parameters, as shown in Figure 2.13.

The relation between the equivalent parameters and the original rough surface morphology parameters is shown in Table 2.2. The deformation model of a single microasperity is shown schematically in Figure 2.14.

The model proposed by M. R. Brake was used to calculate the deformation process of a single microasperity, and the calculation equations are shown in Table 2.3. The mechanical parameters of Ag were used to calculate the deformation process of a microasperity with an equivalent radius $R = 45\,\mu m$, and the results are shown in Figure 2.15.

When the contact area and contact pressure of a single microasperity are obtained based on the aforementioned single microasperity contact theory, the contact area and contact pressure of a rough surface can be obtained using the following integral equation.

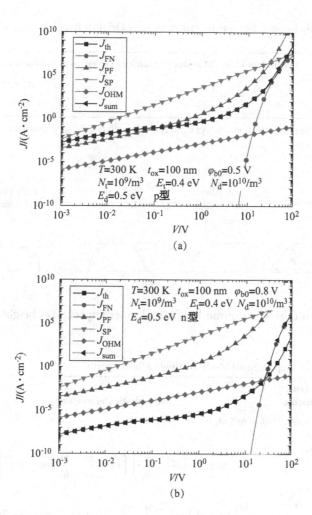

FIGURE 2.12 J-V characteristic curve when the oxide layer is thick and a barrier layer potential is formed. (a) Current density under strong p-type conditions. (b) Current density under n-type conditions.

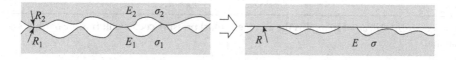

FIGURE 2.13 Schematic diagram of the rough surface contact equivalent GW model.

Contact area of rough surface:

$$A_r(d) = \eta A_n \int_d^{+\infty} \bar{A}(l)\phi(z)dz \qquad (2.38)$$

Contact pressure of rough surface:

$$P(d) = \eta A_n \int_d^{+\infty} \bar{P}(l)\phi(z)dz \qquad (2.39)$$

TABLE 2.2   Relationship between Equivalent Parameters and Original Rough Surface Topography Parameters

Equivalent Radius	Equivalent Elastic Modulus	Standard Deviation of Equivalent Height of Microasperity
$\dfrac{1}{R} = \dfrac{1}{R_1} + \dfrac{1}{R_2}$	$\dfrac{1}{E^*} = \dfrac{1-v_1^2}{E_1} + \dfrac{1-v_2^2}{E_2}$	$\sigma^2 = \sigma_1^2 + \sigma_2^2$

Note: E^* is the equivalent modulus of elasticity; v_1, v_2 are, respectively, Poisson's ratios of the two contact surface materials.

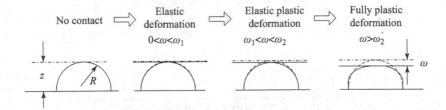

FIGURE 2.14   Schematic deformation model of a single microasperity. (z – height of microasperities.)

TABLE 2.3   Mechanics Computational Model of Single Microasperity

Parameter	Total Elastic Deformation		Elastoplastic Deformation	Total Plastic Deformation
Pressed depth ω	$0 < \omega \leq \omega_1$	$\omega_1 < \omega \leq \omega_2$		$\omega > \omega_2$
Contact area A_a	$\pi R\omega$		$\pi R\left(\omega + \left(\dfrac{\omega-\omega_1}{\omega_2-\omega_1}\right)^2\left(\omega_2 - \dfrac{\omega_2+\omega_1}{\omega_2-\omega_1}(\omega-\omega_2)\right)\right)$	$2\pi R\omega$
Contact pressure P	$kH(\omega/\omega_1)^{1/2}$		$\dfrac{kH}{2\omega_1}(\omega+\omega_1) + \dfrac{H(\omega-\omega_1)^2}{2\omega_1(\omega_2-\omega_1)^3}\big(6\omega_1\omega_2 - 3\omega_1\omega_2 k - 2\omega_2^2 k + \omega_1^2 k - 2\omega_1^2 + (-4\omega_1 + 3\omega_1 k + \omega_2 k)\omega\big)$	H
Contact pressure F	$4/3E^* R^{1/2}\omega^{3/2}$		PA_a	$2\pi RH\omega$

Note: $\omega_1 = (3\pi kH/4E^*)2R$; $\omega_2 = 110\,\omega_1$; k is the average contact pressure coefficient; H is the hardness of the material.

Number of contact microasperity on rough surface:

$$N_c = \eta A_n \int_d^{+\infty} \phi(z)dz \tag{2.40}$$

where η is the surface density of microasperity; A_n is the nominal contact area; d is the average distance between rigid plane and equivalent rough surface.

In this way, when the contacting rough surfaces have different microasperity morphologies, only the contact area $\bar{A}$ and contact pressure $\bar{P}$ of a single microasperity need to be

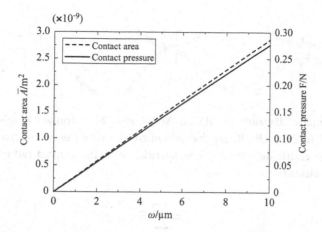

FIGURE 2.15   The deformation process of a single microasperity.

calculated accordingly, and the contact characteristics can be obtained by a similar integral calculation.

The above integration uses a statistical method, which is effective when the number of samples is large and the sample features obey a given distribution. However, for the actual microwave components, since the main role of the component characteristics is often a part of the microasperity at the edge of the contact section, the number of this part of the microasperity is limited, and because part of the microasperity is metal-metal (MM) contact, no nonlinearity is generated, so the number of nonlinear microasperities is even less, when the nonlinearity will have a certain degree of randomness, which cannot be described by the integral method. In addition, if nonideal factors (such as the shape of the microasperity, the distribution of the radius of the microasperity and friction) are considered, the integral method will make the calculation more and more complicated. Therefore, the Monte Carlo method is used here to calculate each microasperity individually, thus making the analysis process simple and allowing the introduction of nonideal factors easily.

2.3.2 Monte Carlo Method Analysis of Microasperity Contact Based on Classical GW Statistical Model

By using the Monte Carlo method, not only the deformation process of each microasperity can be obtained, but also the contact law can be more realistically described in the case of a small number of microasperities. The Monte Carlo method is used to calculate the following procedure: firstly, the microasperities of different heights h_i are randomly generated according to the statistical distribution law of the height distribution of the microasperities on the surface of the actual microwave component; then, the indentation depth ω_i of each microasperity is determined according to the position d_i of the smooth plane, and the pressure F_i and contact area A_i of each microasperity are obtained by analyzing the deformation law of individual microasperities; after that, the forces on each microasperity are summed to obtain the total pressure F of the microwave component surface.

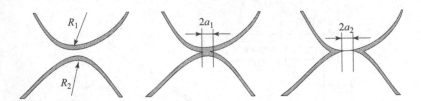

FIGURE 2.16 Schematic diagram of MVM, MOM and MM contact single point structure. (a) MVM. (b) MOM. (c) MM. (R_1, R_2 are the radii of curvature of the two microasperities; a_1 is the contact radius of the elastic deformed microasperities; a_2 is the contact radius of the plastically deformed microasperities.)

$$F = \sum_i F_i \tag{2.41}$$

Dividing the total pressure F by the nominal contact area A_n, we obtain the pressure P as

$$P = \frac{F}{A_n} \tag{2.42}$$

For the contact cross section of microwave components with oxide film or staining on the surface, there are three main types of single-point structure contacts, namely metal-vacuum-metal (MVM) contact, metal-oxide-metal (MOM) contact and metal-metal (MM) contact, as shown in Figure 2.16. Among them, MVM and MOM contacts are the main sources of nonlinear currents, while MM contacts are the main influencing factor of shrinkage resistance, so they are to be counted separately.

When the surfaces of two microwave components are in contact, as the contact pressure increases, for any single microasperity, the MVM contact will change to MOM contact, and the contact pressure will be further increased, the oxide film will break and change into MM contact, as shown in Figure 2.17.

At present, there is no complete theory on the quantitative calculation of oxide film cracking conditions. The book *Electrical Contact Theory and Its Application* describes the qualitative results of early Osias and Trlpp using wax soil plastic asperity model to study the mechanical failure of the film. When an external contact force acts on the asperity, the top asperity material is forced to flow outwards, causing the edge of the dome to be severely bent. The film first ruptures in the circumferential and radial directions at the edge of the dome. At this time, the rupture mainly occurs outside the contact surface and not inside the contact surface; only when the asperity model is further deformed by force, the surface area of the top contact part changes greatly before and after the deformation, and the base metal under the film flows in a large amount, causing the film to break. Therefore, in the description of electrical contact theory, direct metal contact only occurs when the microscopic bumps are severely deformed by force. The above analysis shows that the rupture of the oxide film is directly related to the deformation of the microasperity. Therefore, we use the critical deformation ω_{br} as the rupture condition of the oxide film. When the indentation depth of the microasperity reaches the critical deformation, the oxide film ruptures.

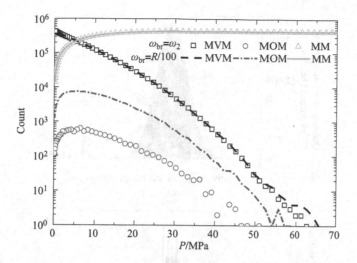

FIGURE 2.17 Variation of the number of MOM contact microasperities with the contact pressure.

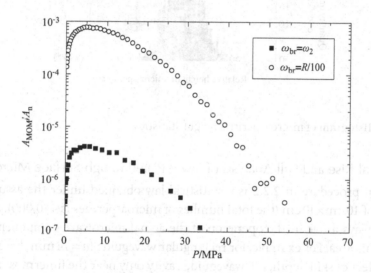

FIGURE 2.18 Variation of MOM contact microasperity area with contact pressure for two cracking conditions of oxide film.

It can be inferred that the larger the radius of the microasperity, the larger the critical deformation. Therefore, assuming that the critical deformation is proportional to R, then

$$\omega_{br} = k_{br} R \tag{2.43}$$

where k_{br} is a constant related to the mechanical properties of metal and oxide film materials and the thickness of the oxide film.

Since the value of k_{br} cannot be determined, we choose two cracking conditions of the oxide film ($\omega_{br}=\omega_2$, $\omega_{br}=R/100$) to conduct Monte Carlo analysis on the contact of the microasperities, as shown in Figure 2.18.

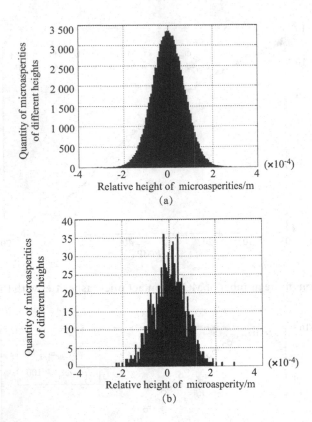

FIGURE 2.19   Histogram of microasperities height statistics.

2.3.3 Statistical Rise and Fall Analysis of Silver-Plated Rough Surface Microcontacts

The calculation procedure in 2.2.2 is a statistical law obtained under the assumption of a nominal area of 10 cm×10 cm (the total number of microasperities is 440,000), and the contact area that plays a role in the operation of the actual microwave components is usually smaller than this area. For example, for rectangular waveguide ($a = 74$ mm, $b = 34$ mm), considering the effect of skin depth, the waveguide cavity only near the innermost microasperities to contribute to the PIM, the total number of microasperities N is estimated to be 981.

Figures 2.19 and 2.20 calculate the relative height distribution of the microasperities and the number of MOM-contacting microasperities when the number of microasperities N is 98,100 and 981, respectively. The standard deviation of the microasperity height is taken as 50 μm, the oxide film cracking condition as $\omega_{br} = R/10$, and the frequency as 2.6 GHz. It can be seen from the figures that there is a significant rise and fall in the number of MOM-contacting microasperities when N decreases.

2.4 NONLINEAR CURRENT THEORY ANALYSIS METHOD FOR METAL CONTACT INTERFACE OF MICROWAVE COMPONENTS

According to the G-W contact model, the flange plate contact of two microwave components can be equivalent to the contact structure of a smooth rigid plane and a rough surface, as shown in Figure 2.21. For the contact structure with a nominal contact area of An,

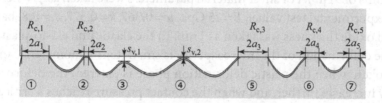

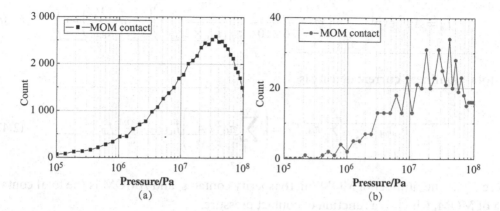

FIGURE 2.20 Variation of between the quantity of MOM contact microasperities and the contact pressure. (a) $N=98100$; (b) $N=981$.

FIGURE 2.21 Schematic diagram of aluminum alloy silver-plated plate contact local. ($S_{v,1}$ and $S_{v,2}$ are the distances between the microasperities and the smooth plane; $a_1 \sim a_5$ are the contact radii of different microasperities, respectively.)

under the contact pressure P, part of the microasperity is not in contact with the smooth plane, such as microasperity ③ and ④ in Figure 2.21; part of the microasperity is in contact with the smooth plane but with small deformation. At this time, the oxide film existing between the microasperity and the smooth plane is MOM contact, such as microasperity ② in Figure 2.21, and the MOM contact microasperity generates nonlinear current and film capacitance; some microasperities contact with smooth planes and undergo severe plastic deformation, and the surface oxide layer is cracked, resulting in MM contact, as shown in Figure 2.21 for microasperities ①, ⑤, ⑥, and ⑦, where MM contact with microasperities generates shrinkage resistance.

For the MOM structure, tunneling current and thermal emission current are mainly considered, and we have

$$J_{nl} = J_{tu} + J_{th} \tag{2.44}$$

where

$$J_{tu} = \frac{6.2 \times 10^{14}}{4t_{ox}^2} \left(\left(\varphi_0 - \frac{V}{2} \right) \exp\left(-1.025 \times 2t_{ox} \left(\varphi_0 - \frac{V}{2} \right)^{1/2} \right) \right.$$

$$\left. - \left(\varphi_0 + \frac{V}{2} \right) \exp\left(-1.025 \times 2t_{ox} \left(\varphi_0 + \frac{V}{2} \right)^{1/2} \right) \right) \tag{2.45}$$

$$J_{th} = AT^2 \exp\left(-\frac{e}{k_B T}\left(\varphi_0 - \frac{e}{16\times10^{-10}\pi\varepsilon_0\varepsilon_{ox}\times2t_{ox}}\right)\right)\left(\exp\left(\frac{eV}{k_B T}\right)-1\right) \quad (2.46)$$

The total nonlinear current density is

$$J_{nl,tot} = \frac{1}{A_n}\sum_{i=1}^{i=N_{co}}\pi a_i^2 J_{nl,i} = \left(\sum_{i=1}^{i=N_{co}}\pi a_i^2 \;/\; A_n\right)J_{nl} = \frac{A_{MOM}}{A_n}J_{nl} \quad (2.47)$$

where N_{co} is the number of MOM microasperity contacts, and AMOM is the total contact area of MOM, which is a function of contact pressure.

Phenomenological morphological parameters of the sample surface were extracted using LSM structure: microasperity decent density of 4.4×107/m², height standard deviation of 7.6 µm and radius of 91 µm. For silver material parameters were taken as: $\rho = 1.65\times10^{-8}\Omega\cdot m$, $H=97$ MPa (experimental test value), $E=75$ Gpa, $\mu=0.367$, $k=0.577$, $\varepsilon_r=8.8$, barrier height $\varphi_0=0.8$eV and oxide thickness was taken as 1 nm. In the elastic and elasto-plastic deformation phase, the deformation of the microasperity is small (tens nanometers), and the oxide film is not broken; when the plastic deformation phase is entered, the deformation of the microasperity increases further, and when the contact pressure reaches a critical value, the oxide film breaks, and the microasperity contact is transformed from an MOM contact to an MM contact. The critical value is related to the shape of the microasperity and the mechanical properties of the material, and the critical condition of elastic-plastic deformation and plastic deformation is adopted here as the condition of oxide film rupture.

Figure 2.22 gives the number of microasperities in various deformation stages at different contact pressures. It can be seen from the figure that the number of MM-contacting microasperities accounts for the majority; with the increase of pressure, all microasperities are eventually transformed into MM-contacting; with the increase of pressure, the number of MOM-contacting microasperities first increases and then decreases.

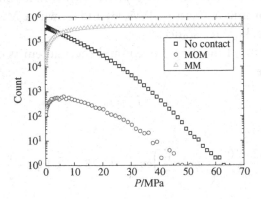

FIGURE 2.22 Number of microasperities in different deformation stages at different contact pressures.

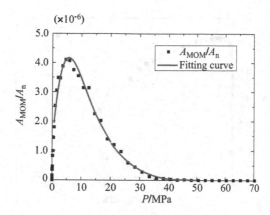

FIGURE 2.23 The ratio of MOM contact area to nominal contact area varies with contact pressure.

TABLE 2.4 The Values of the Fitting Coefficients C_i

Pressure/Mpa	C_0	C_1	C_2	C_3	C_4	C_5
1~20	$1.267\,9\times10^{-6}$	$1.243\,8\times10^{-12}$	$-1.740\,4\times10^{-19}$	$8.937\,1\times10^{-27}$	$-1.857\,7\times10^{-34}$	$9.390\,5\times10^{-43}$
20~50	$1.099\,6\times10^{-5}$	$-1.104\,1\times10^{-12}$	$5.040\,2\times10^{-20}$	$-1.283\,7\times10^{-27}$	$1.747\,4\times10^{-35}$	$-9.775\,7\times10^{-44}$

The ratio of the MOM contact area A_{MOM} to the nominal contact area A_n is a function of the contact pressure. The Maclaurin expansion of this function is carried out and the higher-order terms are ignored. We have

$$\frac{A_{\text{MOM}}(P)}{A_n} \approx \sum_{i=1}^{7} c_i P^i \qquad (2.48)$$

This polynomial is used to fit Figure 2.23 to get the coefficients of each order, as shown in Table 2.4.

Expanding the polynomial of the nonlinear J-V characteristic, we have

$$J_{nl,tot} = \frac{A_{\text{MOM}}}{A_n} J_{nl} \approx \sum_{m=1}^{7} a_m V^m \qquad (2.49)$$

where a_m is polynomial coefficient.

The results of the calculation for $a3$, $a5$和$a7$ are shown in Figure 2.24.

The relationship curves of each secondary term nonlinear current with the variation of contact pressure are given in Figure 2.25. From the figure, it can be seen that the nonlinear current of each secondary term increases and then decreases with the increase of the contact pressure.

For the gold material surface contact, considering the existence of adsorption film on the gold surface, the contact situation of the microasperity is similar to that of the silver plate surface contact. The electrical and mechanical parameters of the gold used were

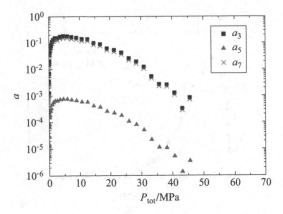

FIGURE 2.24 Calculated values of a_3, a_5 and a_7 for aluminum alloy silver-plated plate specimens at different pressures.

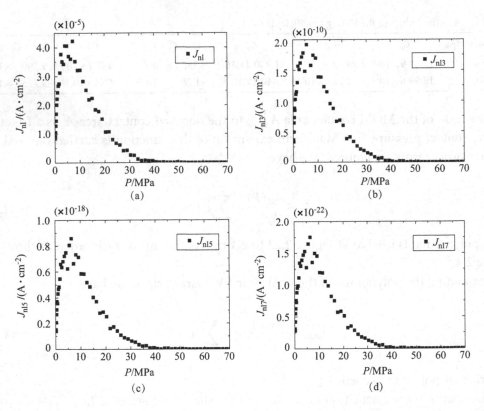

FIGURE 2.25 The variation curve of nonlinear current with contact pressure for each secondary term of aluminum alloy silver-plated flat surface contact. (a) 1st order nonlinear current. (b) 3rd order nonlinear currents. (c) 5th order nonlinear currents. (d) 7th order nonlinear currents.

calculated as $\rho = 2.3 \times 10^{-8}\Omega \cdot cm$, $H = 200\,MPa$, $E = 84$ CPa, $\mu = 0.42$, $k = 0.577$. Figure 2.26 gives the comparison of gold-plated and silver-plated surface contact nonlinear currents under the same surface morphology parameters. Due to the difference of potential barriers

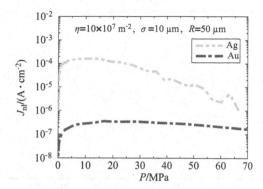

FIGURE 2.26 The comparison of the nonlinear current contact between gold-plated and silver-plated surfaces under the same surface topography parameters.

between gold-plated and silver-plated surfaces, the surface contact nonlinear currents of gold-plated samples are smaller. Therefore, in order to obtain lower PIM characteristics, for actual microwave components, gold-plated materials can be considered.

2.5 ANALYTICAL DERIVATION OF NONLINEAR CURRENT IN CONTACT OF PARTS

A single microasperity is in contact with a rigid smooth plane, and with the increase of contact pressure, the microasperity is gradually deformed. The deformation process of the microasperity is divided into three stages, i.e., total elastic deformation, elastoplastic deformation and fully plastic deformation. The mechanical models of each deformation stage are shown in Table 2.3.

The surface of the actual microwave component consists of many microasperities, the heights of which obey the Gaussian distribution, so we have

$$\phi(z) = \frac{1}{\sqrt{2\pi}\sigma} \exp\left(-\frac{z^2}{2\sigma^2}\right) \tag{2.50}$$

As seen in Figure 2.27, the vast majority of the contacting microasperities are in the fully plastic deformation stage.

The total contact pressure is the sum of the forces of each microasperity. Since the number of microasperities in the complete plastic deformation stage accounts for the majority, the contribution of other types of contact to the total contact pressure can be ignored, namely

$$F_{tot} = 2\pi R H \sum_i \omega_i = 2\pi R H N_{tot} \int_{d+\omega_2}^{\infty} (z-d)\phi(z)dz$$

$$= 2\pi R H N_{tot}\left(\frac{\sigma}{\sqrt{2\pi}}\exp\left(-\frac{(d+\omega_2)^2}{2\sigma^2}\right) - \frac{d}{2}\mathrm{erfc}\left(\frac{d+\omega_2}{\sigma\sqrt{2}}\right)\right) \tag{2.51}$$

where F_{tot} is the total contact pressure, and N_{tot} is the total number of microasperities.

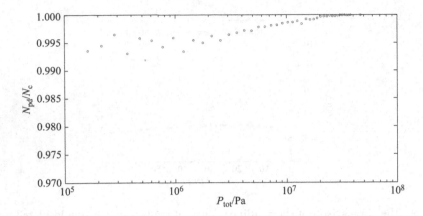

FIGURE 2.27 The variation curve of the ratio of the number of microasperities N_{pd} to the total number of contact microasperities N_c with the total contact pressure P_{tot} in the complete plastic deformation stage.

The total contact pressure is

$$P_{tot} = \frac{F_{tot}}{A_n} = 2\pi RH \frac{N_{tot}}{A_n}\left(\frac{\sigma}{\sqrt{2\pi}}\exp\left(-\frac{(d+\omega_2)^2}{2\sigma^2}\right) - \frac{d}{2}\text{erfc}\left(\frac{d+\omega_2}{\sigma\sqrt{2}}\right)\right)$$

$$= 2\pi RH\eta\left(\frac{\sigma}{\sqrt{2\pi}}\exp\left(-\frac{(d+\omega_2)^2}{2\sigma^2}\right) - \frac{d}{2}\text{erfc}\left(\frac{d+\omega_2}{\sigma\sqrt{2}}\right)\right) \quad (2.52)$$

The contact area of the MOM consists of the sum of the contact areas of the fully elastic deformed and elastoplastic deformed microasperities. It is assumed that the contact area of each microasperity at each stage of deformation can be expressed as

$$A_i = 2\pi R\omega_i = 2\pi R(z_i - d) \quad (2.53)$$

Since z_i obeys a Gaussian distribution $\varphi(z)$, the distribution function of the area of each contacting microasperities can be expressed as

$$F_A(y) = P\left\{d \leq z \leq \frac{y}{2\pi R} + d\right\}$$

$$= \int_{-\infty}^{\frac{y}{2\pi R}+d} \phi(z)dz - \int_{-\infty}^{d} \phi(z)dz \quad (2.54)$$

Then the total contact area of MOM is

$$A_{\S P\S d\S P} = \sum_{A, \leq 2\pi\beta\omega_2} A_i = \frac{PA_n}{H}F_A(2\pi R\omega_2)$$

$$= \frac{\pi PA_n}{2H}\left(\text{erf}\left(\frac{\omega_2+d}{\sigma\sqrt{2}}\right) - \text{erf}\left(\frac{d}{\sigma\sqrt{2}}\right)\right) \quad (2.55)$$

The total nonlinear surface current is

$$I(P,V_a) = A_{\text{MoM}} J_{\text{MOM}}$$

$$= A_{\text{MOM}} \left(\left(6.2 \times 10^{14} / t_{ox}^2 \right) \left(\left(\varphi_0 - \frac{V_a}{2} \right) \exp\left(-1.025 t_{ox} \left(\varphi_0 - \frac{V_a}{2} \right)^{1/2} \right) \right. \right.$$

$$\left. - \left(\varphi_0 + \frac{V_a}{2} \right) \exp\left(-1.025 t_{ox} \left(\varphi_0 + \frac{V_a}{2} \right)^{1/2} \right) \right)$$

$$\left. + AT^2 \exp\left(-\frac{e}{k_B T} \left(\varphi_0 - \frac{e}{16 \times 10^{-10} \pi \varepsilon_0 \varepsilon_{ox} t_{ox}} \right) \right) \left(\exp\left(\frac{eV_a}{k_B T} \right) - 1 \right) \right) \quad (2.56)$$

where J_{MOM} is the current density of MOM.

The total nonlinear surface current density is

$$J(P,V_a) = \frac{A_{\text{MOM}}}{A_n} J_{\text{MOM}}$$

$$= \frac{A_{\text{MOM}}}{A_n} \left(\left(6.2 \times 10^{14} / t_{ox}^2 \right) \left(\left(\varphi_0 - \frac{V_a}{2} \right) \exp\left(-1.025 t_{ox} \left(\varphi_0 - \frac{V_a}{2} \right)^{1/2} \right) \right. \right.$$

$$\left. - \left(\varphi_0 + \frac{V_a}{2} \right) \exp\left(-1.025 t_{ox} \left(\varphi_0 + \frac{V_a}{2} \right)^{1/2} \right) \right)$$

$$\left. + AT^2 \exp\left(-\frac{e}{k_B T} \left(\varphi_0 - \frac{e}{16 \times 10^{-10} \pi \varepsilon_0 \varepsilon_0 t_{ax}} \right) \right) \left(\exp\left(\frac{eV_a}{k_p T} \right) - 1 \right) \right) \quad (2.57)$$

where

$$\frac{A_{\S P \S \dot{a} \S P}}{A_n} = \frac{\pi P}{2H} \left(\text{erf}\left(\frac{\omega_2 + d}{\sigma \sqrt{2}} \right) - \text{erf}\left(\frac{d}{\sigma \sqrt{2}} \right) \right) \quad (2.58)$$

$$d = f^{-1}(P) \quad (2.59)$$

$$f(d) = 2\pi R H \eta \left(\frac{\sigma}{\sqrt{2\pi}} \exp\left(-\frac{(d + \omega_2)^2}{2\sigma^2} \right) - \frac{d}{2} \text{erfc}\left(\frac{d + \omega_2}{\sigma \sqrt{2}} \right) \right) \quad (2.60)$$

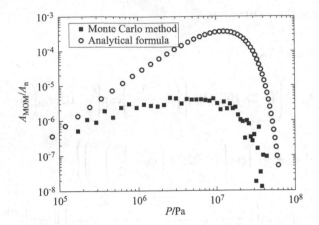

FIGURE 2.28 The ratio of A_{MOM} and A_n vs. the contact pressure P.

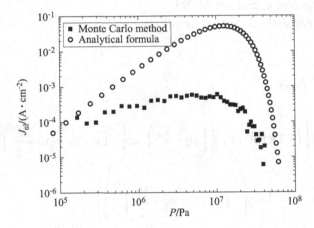

FIGURE 2.29 Total nonlinear current density J_{nl} vs. contact pressure P (voltage V is taken as 1 mV).

The curve of the variation of the ratio of $AMOM$ to A_n with the contact pressure obtained from the calculation of Equation (2.58) is shown in Figure 2.28. From the figure, it can be seen that the ratio results obtained using Equation (2.58) are larger than those obtained using the Monte Carlo method. This is due to the fact that the assumption of Equation (2.55) for each contact area value is larger than the actual value, so that the final total A_{MOM} to A_n ratio is obtained to be larger. The total nonlinear current density J_{nl} versus contact pressure P is shown in Figure 2.29.

2.6 ANALYSIS OF ELECTROTHERMAL COUPLING MECHANISM AND ANALYSIS OF PASSIVE INTERMODULATION PRODUCTS MODELING

In addition to contact nonlinearity, material nonlinearity also generates PIM. One of the key points in the study of material nonlinearity is the electrothermal coupling effect. This section will introduce the electrothermal coupling effect and the mechanism of PIM generation in microwave components based on electrothermal coupling.

2.6.1 Electrothermal Coupling Effect

For a metallic conductor, the current passing through it generates its own heat, which further leads to an increase in its own temperature, and the temperature increase leads to an increase in resistance, which is the electrothermal coupling effect. Due to the electro-thermal coupling effect, in the case of direct current, the current-voltage relationship of the conductor itself has nonlinear characteristics. To demonstrate the characteristics of the electrothermal coupling effect, molybdenum wire is selected to measure its current-voltage relationship in the case of direct current. Molybdenum wire is the base material of woven grid antenna, which has small radius, large DC resistance and strong electrothermal cou-pling effect. Figure 2.30 shows the current-voltage relationship for a gold-plated molybde-num wire with a length of 10 cm. There are two types of molybdenum wires, one with gold plating on the surface and the other with gold plating on the surface followed by dielectric plating. The DC test equipment is Agilent 2902A. Key test option settings: speed: dielectric, scanning 0~2V takes about 2.34s.

It can be seen from Figure 2.30b that as the voltage increases from 0 to 2.0 V, the current also increases from 0 to 0.4A. In the 0–0.35 V low voltage region, the current-voltage rela-tionship is linear, and the resistance of the molybdenum wire in this region is calculated to be 4Ω; while in the region where the voltage is greater than 0.5 V, the current-voltage relationship deviates significantly from linearity, which is different from the nonlinear current-voltage curve of the diode. Different from the nonlinear current-voltage curve characteristics of the diode, the slope of the current-voltage curve of the molybdenum wire decreases with the increase of the voltage, which indicates that the resistance of the molybdenum wire increases with the increase of the voltage. When the voltage increases, the current through the molybdenum wire also increases, which will lead to the increase of its temperature and further lead to the increase of resistance. Therefore, the slope of the current-voltage curve in Figure 2.30 decreases with the increase of voltage, that is, the resistance increases with the increase of voltage. Although the current-voltage curve is nonlinear, the conductivity characteristics at each moment are ohmic under each voltage. The nonlinearity of the curve comes from the increase in the temperature of the accumu-lated molybdenum wire over time. However, the temperature of the molybdenum wire

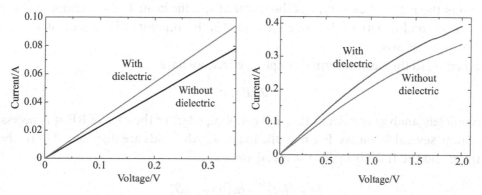

FIGURE 2.30 Current-voltage relationship of 10 cm long gold-plated molybdenum wire. (a) 0~0.35 V interval. (b) 0~2.0 V interval.

cannot rise indefinitely. If the applied voltage is fixed, then as time increases, the heat exchange between the molybdenum wire and the surrounding air will eventually reach equilibrium, at which point the temperature of the wire and the resistance will no longer change. Therefore, this type of curve is closely related to the scanning process, mainly related to the duration of the applied voltage, and also related to the thermal environment (such as thermal conductivity, heat exchange coefficient).

2.6.2 Electrothermal Coupling Effect and Passive Intermodulation Generation Mechanism under High-Frequency Field

The electrothermal coupling effect in the DC case was described earlier. For the microwave case, the surface current flows in the range of several skin depths on the conductor surface, which can be calculated through the following equation.

$$\delta = \sqrt{\frac{1}{\pi f \mu_0 \sigma}} \tag{2.61}$$

where f is the frequency of the RF current; μ_0 is the material permeability; and σ is the material conductivity.

According to Maxwell's equation, we have

$$\begin{cases} \nabla \times \boldsymbol{E} = -\mu \dfrac{\partial \boldsymbol{H}}{\partial t} \\ \nabla \times \boldsymbol{H} = \boldsymbol{J} + \varepsilon \dfrac{\partial \boldsymbol{E}}{\partial t} = \sigma \boldsymbol{E} + \varepsilon \dfrac{\partial \boldsymbol{E}}{\partial t} \end{cases} \tag{2.62}$$

Since the current density J changes with time, the heat generated changes with time t, resulting in the electrical conductivity σ changing with time, and after the electrical conductivity interacts with the carrier electric field, a new frequency product will appear. For thermal property, we have

$$\rho c \frac{\partial T}{\partial t} + \nabla \cdot (-\lambda \nabla T) = Q \tag{2.63}$$

where ρ is the material density; c is the material specific heat; T is the temperature; Λ is the thermal conductivity of the material; and Q is the amount of heat generated per unit volume, i.e., heat loss.

According to the electrothermal coupling effect, we have

$$Q = \boldsymbol{J} \cdot \boldsymbol{E} = \sigma |E|^2 \tag{2.64}$$

To completely analyze the electrothermal coupling effect in the case of RF, it is necessary to combine several formulas. For metallic materials, the fields are distributed in the body, and the E and the H fields inside the metal are

$$\begin{cases} \boldsymbol{H} = H_0 e^{-r/\delta} \cos(\omega t - r/\delta) \\ \boldsymbol{E} = \sqrt{\dfrac{\mu \omega}{\delta}} H_0 e^{-r/\delta} \cos\left(\omega t - r/\delta + \dfrac{\pi}{4}\right) \end{cases} \tag{2.65}$$

where H_0 is magnetic field strength at the surface.

For the single-carrier case, the surface currents are approximated as distributed within a skin depth, and the electric field and the currents differ by only one conductivity between them, so the electric field can be treated similarly under the assumption that the electric field exists only within a skin depth. For the dual-carrier case with frequencies ω_1 and ω_2, it is assumed that the amplitudes are equal and the phases are the same, and we have

$$Q = \mathbf{J} \cdot \mathbf{E} = \sigma (E_1 + E_2)^2$$

$$= \sigma \left(\sqrt{\frac{\mu \omega_1}{\delta_1}} H_0 e^{-t/\delta_1} \cos\left(\omega_1 t - r/\delta_1 + \frac{\pi}{4} \right) + \sqrt{\frac{\mu \omega_2}{\delta_2}} H_0 e^{-r/\delta_2} \cos\left(\omega_2 t - r/\delta_2 + \frac{\pi}{4} \right) \right)^2$$

$$= \frac{\mu \omega_1}{2} H_0^2 e^{-2r/\delta_1} + \frac{\mu \omega_1}{2} H_0^2 e^{-2r/\delta_1} \cos\left(2\omega_1 t - 2r/\delta_1 + \frac{\pi}{2} \right)$$

$$+ \frac{\mu \omega_2}{2} H_0^2 e^{-2r/\delta_2} + \frac{\mu \omega_2}{2} H_0^2 e^{-2r/\delta_2} \cos\left(2\omega_2 t - 2r/\delta_2 + \frac{\pi}{2} \right)$$

$$+ \mu\sqrt{\omega_1 \omega_2} H_0^2 e^{-r/\delta_1 - r/\delta_2} \cos\left((\omega_1 + \omega_2)t - r/\delta_1 - r/\delta_2 + \frac{\pi}{2} \right)$$

$$+ \mu\sqrt{\omega_1 \omega_2} H_0^2 e^{-r/\delta_1 - r/\delta_2} \cos\left((\omega_1 - \omega_2)t - r/\delta_1 + r/\delta_2 \right) \qquad (2.66)$$

where E_1 and E_2 are electric fields of two carriers in the metal; δ_1 and δ_2 are skin depths of the two carrier electric fields, respectively.

It can be seen from Equation 2.66 that the heat generated in the metal due to heat loss has both a DC part and an AC part in the case of dual carrier. The heat of the DC part will cause the metal temperature to rise from the initial temperature to the equilibrium temperature; the AC part will cause the metal temperature to rise from the initial temperature to the "equilibrium" temperature, and the temperature will oscillate during the rising process and after the equilibrium. In Equation (2.66), the frequency components of heat are $2\omega_1$, $2\omega_2$, $\omega_1 + \omega_2$ and $\omega_1 - \omega_2$. For radio frequency and microwave, the frequency is up to hundreds or thousands of MHz, and the thermal response of the material has low-pass characteristics, as shown in Figure 2.31, so the high frequency terms $2\omega_1$, $2\omega_2$ and $\omega_1 + \omega_2$ can be ignored. In addition, when the heat frequency of the dual carrier frequency is not very large, it can be approximated as $\delta_1 = \delta_2$, so Equation 2.66 can be approximated as

$$Q \approx \frac{\mu \omega_1}{2} H_0^2 e^{-2r/\delta_1} + \frac{\mu \omega_2}{2} H_0^2 e^{-2r/\delta_2} + \mu\sqrt{\omega_1 \omega_2} H_0^2 e^{-2r/\delta_1} \cos(\omega_1 - \omega_2)t \qquad (2.67)$$

Although it is difficult to jointly solve Equations 2.63 and 2.67 directly to obtain an analytical solution, some qualitative results can be obtained by analysis. After reaching equilibrium, the temperature of the metal oscillates with frequency $\omega_1 - \omega_2$, so the temperature T can be expressed as

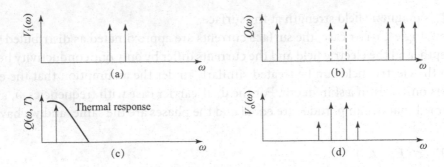

FIGURE 2.31   Schematic diagram of electrothermal coupling process. (a) The frequency spectrum of the dual carrier. (b) The heat loss spectrum of the dual carrier. (c) The frequency component of heat loss that can be coupled into the thermal response process. (d) The spectrum after electrothermal coupling.

$$T(t) = T_0 + A\sin(\omega_1 - \omega_2)t \tag{2.68}$$

where A is the coefficient.

The conductivity σ of metal decreases with increasing temperature, and the change in conductivity with temperature can be expanded into the following polynomial:

$$\sigma(T) = \sigma_0 \left(1 + BT + CT^2 + \cdots\right) \tag{2.69}$$

where B and C are the coefficients. The higher the power, the smaller the coefficient, so it will only expand to the second power in the following calculations.

In the case of dual carrier, the product of the time-varying conductivity and the dual carrier electric field is

$$
\begin{aligned}
\sigma(T)E &= \sigma_0\left(1 + BT + CT^2\right)\left(E\sin(\omega_1 t) + E\sin(\omega_2 t)\right) \\
&= \sigma_0\left(1 + B\left(T_0 + A\sin\left((\omega_1 - \omega_2)t\right)\right)\right)\left(E\sin(\omega_1 t) + E\sin(\omega_2 t)\right) \\
&\quad + \sigma_0 C\left(T_0 + A\sin\left((\omega_1 - \omega_2)t\right)\right)^2 \left(E\sin(\omega_1 t) + E\sin(\omega_2 t)\right) \\
&= \sigma_0\left(1 + BT_0 + CT_0^2 + \frac{1}{2}CA^2\right)E\left(\sin(\omega_1 t) + \sin(\omega_2 t)\right) \\
&\quad - \sigma_0 A\left(\frac{1}{2}B + CT_0\right)E\left(\cos\left((2\omega_1 - \omega_2)t\right) - \cos\left((2\omega_2 - \omega_1)t\right) + \cos(\omega_1 t)\right. \\
&\quad \left. - \cos(\omega_2 t)\right) - \frac{1}{4}\sigma_0 CA^2 E\left(\sin\left((2\omega_2 - \omega_1)t\right) + \sin\left((2\omega_1 - \omega_2)t\right)\right) \\
&\quad - \frac{1}{4}\sigma_0 CA^2 E\left(\sin\left((3\omega_1 - 2\omega_2)t\right) + \sin\left((3\omega_2 - 2\omega_1)t\right)\right) \tag{2.70}
\end{aligned}
$$

It can be clearly seen from Equation (2.70) that after the time-varying conductivity is multiplied by the electric field of the dual carrier, an electric field with a new frequency appears. The new frequencies are $2\omega_1 - \omega_2$, $2\omega_2 - \omega_1$, $3\omega_1 - 2\omega_2$ and $3\omega_2 - 2\omega_1$, namely the 3rd and 5th order PIM components; the size of each order PIM component is related to the temperature rise, the amplitude and the expansion coefficient of the conductivity with the temperature. In addition, it can be seen that the magnitude of the PIM component is proportional to $\sigma_0 E$ (that is, the current density in the metal). For the same structure, σ_0 depends on the properties of the material. The larger the power of the carrier, the larger the PIM component (that is, the power of the PIM product).

2.7 SUMMARY

Internationally, for the PIM nonlinear mechanism of the metal contact structure of microwave components, the accurate physical parameters have not been established, but the tunneling current and thermal emission current characteristic formulas of the MIM (Metal-Insulation-Metal) structure are simply quoted. Combined with surface topography and composition analysis, this chapter determines the physical parameters for calculating the metal contact junction of microwave components; based on the study of the main carrier nonlinear transport processes such as quantum tunneling, thermionic emission, grain boundary transport and narrow gap transport, this chapter preliminarily reveals the main nonlinear transport physical mechanisms of Ag-Ag metal contact junctions and establishes the nonlinear current density equation for microasperity contacts. In this chapter, the nonlinear current characteristic equation of single point structure is combined with the microwave component contact junction microasperity contact Monte Carlo method to realize the statistical analysis method of nonlinear surface current of contact interface, and the nonlinear surface current equation is established accordingly to obtain the analytical model of metal contact nonlinearity. In addition, this chapter also introduces the electro-thermal coupling effect, and analyzes the mechanism of PIM based on the electrothermal coupling effect. Because the mechanism of PIM is very complex and there are many factors affecting it, this chapter only analyzes the PIM caused by contact nonlinearity and electrothermal coupling. In order to facilitate the analysis and calculation, some assumptions are also used in the research process. These approximations or assumptions can not fully reflect the real PIM situation in microwave components. Therefore, the approximate conditions of these assumptions will be further studied in order to establish a more accurate PIM physical model.

BIBLIOGRAPHY

1. Lui P.L. Passive intermodulation interference in communication systems. *Electronics & Communication Engineering Journal*, 1990, 2: 109–118.
2. Zhang S.Q., Fu D. M., Ge D. B. The effects of passive intermodulation interference on the anti-noise property of communications systems. *Chinese Journal of Radio Science*, 2002, 17(2): 138–142.
3. Boyhan J. W., Lenzinc H. F., Koduru C. Satellite passive intermodulation: Systems considerations. *IEEE Transactions on Aerospace and Electronic Systems*, 1996, 32 (3): 1058–1064.

4. Wang H. N., Liang J. G., Wang J. Q., et al. Review of passive intermodulation in HPM condition. *Journal of Microwaves*, 2005, 21: 1–6.

5. Zhang S. Q., Ge D. B. Intermodulation interference due to passive nonlinearity in communication systems. *Journal of Shaanxi Normal University (Natural Science Edition)*, 2004, 32(1): 58–62.

6. Liu E. K., Zhu B. S., Luo J. S. *The Physics of Semiconductors* (7th Editor). Beijing: Publishing House of Electronics Industry, 2011.

7. Popov V. *Contact Mechanics and Friction: Physical Principles and Applications*. Berlin: Springer Science & Business Media, 2010.

8. Valentin L. P. *Contacts Mechenics and Friction Physical Principes and Applications*. Beijing: Tsinghua University Press, 2011.

9. Slade P. G. *Electrical Contacts: Principles and Applications*. Abingdon: CRC Press Taylor & Francis Croup, 1999.

10. Braunovic M., Konchits V. V., Myshkin N. K. *Electrical Contacts: Fun-Damentals, Applications and Technology*. Boca Raton, FL: CRC Press, 2007.

11. Cheng L. C. *Electrical Contacts Fundamentals and Applications*. Beijing: China Machine Press, 1988.

12. Simmons J. G. Ceneralized formula for the electric tunnel effect between similar electrodes separated by a thin insulating film. *Journal of Applied Physics*, 1963, 34 (6): 1793–1803.

13. Simmons J. G. Potential barriers and emission - limited current flow between closely spaced parallel metal electrodes. *Journal of Applied Physics*, 1964, 35 (8): 2472–2481.

14. Zhou Y. C. *Physical Mechanics Frontier*. Beijing: Science Press, 2018.

15. Greenwood J. A., Williamson J. B. P. Contact of nominally flat surfaces. *Proceedings Of The Royal Society A Mathematical Physical & Engineering Sciences*, 1966, 295 (1442): 300–319.

16. Brake M. R. An analytical elastic - perfectly plastic contact model. *International Journal of Solids & Structures*, 2012, 49 (22): 3129–3141.

17. Johnson K. L. Contact mechanics. *Journal of Tribology*, 1985, 108 (4): 464.

18. Blau P. J. *Friction Science and Technology: From Concepts to Applications*. Abingdon: CRC Press Taylor & Francis Group, 2009.

19. Whitley J. H. Concerning normal force requirements for precious metal plated connectors. *Proceedings of the 20 TH Annual Connector and Interconnect Technology Symposium*, Philadelphia, PA, 1987.

20. Kantner E. A., Hobgood L. D. Hertz stress as an indicator of connector reliability. *Connection Technology*, 1989, 5 (3): 14–22.

21. Ye M. *Research on Passive Intermodulation Mechanism of MIM Structure in Microwave Components*. Xi'An: Xi'An Jiaotong University, 2010.

22. Vicente C., Hartnagel H. L. Passive-intermodulation analysis between rough rectangular waveguide flanges. *IEEE Transactions on Microwave Theory and Techniques*, 2005, 53 (8): 2515–2525.

23. Zhao X., He Y., Ye M., et al. Analytic passive intermodulation model for flange connection based on metallic contact nonlinearity approximation. *IEEE Transactions on Microwave Theory and Techniques*, 2017, 65 (5): 2279–2287.

24. Vicente C., Wolk D., Hartnagel H. L., et al. Experimental analysis of passive intermodulation at waveguide flange bolted connections. *IEEE Transactions on Microwave Theory & Techniques*, 2007, 55 (5): 1018–1028.

25. Wu D. W., Xie Y. J., Kuang Y., et al. Prediction of passive intermodulation on grid reflector antenna using collaborative simulation: Multiscale equivalent method and nonlinear model. *IEEE Transactions Antennas and Propagation*, 2018, 66 (3): 1516–1521.

26. You J. W., Zhang J. F., Gu W. H., et al. Numerical analysis of passive inter-modulation arisen from nonlinear contacts in HPMW devices. *IEEE Transactions on Electromagnetic Compatibility*, 2018, 60 (5): 1470–1480.

27. Zhang K., Li T. J., Jiang J. Passive intermodulation of contact nonlinearity on microwave connectors. *IEEE Transactions on Electromagnetic Compatibility*, 2018, 60 (2): 513–519.
28. Wilkerson, J. R., et al. Electro - thermal theory of intermodulation distortion in lossy microwave components. *IEEE Transactions on Microwave Theory and Techniques*, 2008, 56 (12): 2717–2725.
29. He Y., Wang Q., Hu T. C., et al. Calculation model for thermal-caused passive intermodulation product of microstrip lines. *Journal of Xidian University*, 2017, 44: 120–126.
30. Wilcox J. Z., Molmud P. Thermal heating contribution to intermodulation fields in coaxial waveguides. *IEEE Transactions on Communications*, 1976, 24 (2): 238–243.
31. Rocas E., Collado C., Orloff N. D., et al. Passive intermodulation due to self - heating in printed transmission lines. *IEEE Transactions on Microwave Theory and Techniques*, 2011, 59 (2): 311–322.
32. Ye M., He Y. N., Cui W. Z. Passive intermodulation mechanism of microstrip lines based on the electro-thermal coupling effect. *Chinese Journal of Radio Science*, 2013, 28(2):1–5.
33. Wilkerson J. R., Lam P. G., et al. Distributed passive intermodulation distortion on transmission lines. *IEEE Transactions on Microwave Theory and Techniques*, 2011, 59 (5): 1190–1205.

CHAPTER 3

Passive Intermodulation Analysis and Evaluation Techniques for Microwave Components

Rui Wang, He Bai, Jianfeng Zhang,
Jiangtao Huangfu, and Yongjun Xie

CONTENTS

DOI: 10.1201/9781003269953-3

3.1 GENERAL OVERVIEW

With the development of modern communication systems to high-power, wide-bandwidth and high sensitivity, the impact of PIM on system performance will become more and more serious, and may even lead to the collapse of the entire system. Therefore, it is imperative to study the potential PIM sources in microwave systems and to predict the magnitude of their PIM products. From the many factors that produce PIM, methods and measures can be found to effectively reduce and control PIM hazards, thus providing a theoretical and practical basis for the development and production of high-power microwave systems with low PIM.

The calculation of PIM products has always been a basic theory and engineering problem in communication engineering. In theory, a reasonable physical model is established based on the generation mechanism of PIM, and the power of PIM products is calculated mathematically. Initially, Sarkozy used a method of nth order polynomial to fit a slight nonlinearity to predict PIM interference. He also made a software package to include the band-limiting effect with Gaussian spectral density, and the results of the analysis of 12 channels in communication satellites were very consistent with the actual measurements. Chapman et al. did a more comprehensive PIM research, including generation mechanism, analytical calculation and detection methods. In 1981, Eng et al. made a theoretical analysis of the order and type of PIM, and proposed the use of the commonly used tree search algorithm to analyze. Boyhan, on the other hand, derived a formula for the calculation of higher-order PIM products from Sunde's basic equation. Later, a large number of researchers developed equivalent circuit models for some specific structures to analyze their nonlinear effects. For example, Vicente et al. studied the PIM interference problem at waveguide flange connections based on the establishment of an equivalent circuit model and analysis of the electron tunneling effect, while Zelenchuk et al. analyzed the PIM characteristics of the microstrip lines printed on circuit boards using a transmission line model with terminated nonlinear parasitic resistance.

It is relatively difficult to analyze PIM products by numerical analysis methods. The traditional numerical analysis methods include Finite Difference Time Domain (FDTD) method and Finite Element Method (FEM). In the paper issued by Ishibashi et al., the current-type nonlinear PIM was analyzed using the FDTD method, which assumes that the resistivity in the simulation region is nonlinear and introduces the third-order nonlinear resistivity coefficients into Maxwell's equations, followed by the electromagnetic simulation using the classical FDTD method, which uses open coaxial lines to verify the simulation results. Such traditional numerical simulation methods often have disadvantages such as high computing cost and slow speed. Bolli and Selleri et al. proposed a scheme to analyze PIM products using the Time-Domain Physical Optic (TDPO) method. This method requires a small number of variables and is particularly suitable for PIM analysis of electrically large-size antennas. Since the mutual coupling effect between points on the surface of the object is not considered, this method can also avoid many errors. However, the method also has many weak points; for example, the electromagnetic wave wavelength of the simulation excitation source must be much smaller than the size of the scatterer (radius of curvature); the simulation results are only reliable on the specular reflection path, and the results in other directions do not satisfy the reciprocity theorem.

3.2 PASSIVE INTERMODULATION NONLINEAR ANALYSIS METHOD

The calculation of PIM products has always been a basic theory and engineering problem in communication engineering. In the 1930s and 1940s, Bennet et al. used the double Fourier series expansion method to calculate and analyze the case of two sinusoidal signals passing through a half-wavelength linear rectifier and extended it to the propagation function of the rectifier and expressed the conducting region as an nth-order polynomial. Later, Feuerstein performed a similar analysis on the multicarrier case; Brockbawk et al. analyzed the multicarrier case in a coaxial cable telephone system using the power series of nonlinear functions, and Middleton conducted a classical analysis of random noise of nonlinear devices. These methods are general for both active and passive intermodulation. However, in view of the special nature of PIM, researchers have made many new theoretical contributions.

Shiquan Zhang used the Fourier series method to derive the general expression and its basic characteristics of PIM product in the two-carrier case and numerically calculated the PIM product of two types of transfer functions; the synthetic interference model hypothesis and characteristic function method is used to simplify the PIM problem. The statistical characteristics of the total interference including PIM interference are mathematically analyzed, and then the influence of PIM on the antinoise performance of the communication system is numerically simulated; the power series method is used to derive the polynomial expressions for predicting the amplitude and power of higher-order PIM (especially odd intermodulation) products from lower-order PIM measurements and the corresponding matrix expressions, and the programming is carried out to achieve the power prediction of 5th order PIM from 3rd order PIM measurements. In addition, comparison with the experimental value is performed initially to confirm the correctness and validity of this method; he also conducts an example analysis of mobile communication and satellite communication at the system level. Besides, Haining Wang et al. used an inverted microscope to calculate the PIM products; Congmin Wang et al. used the power series method to analyze the 5th order PIM products produced by the aluminum-aluminum oxide-aluminum junction in the antenna.

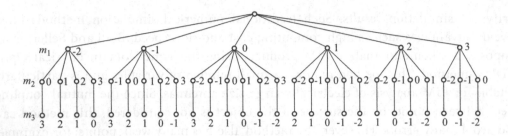

FIGURE 3.1 Structure of the search tree algorithm adopted.

In recent years, researchers have achieved many theoretical research results, both in mathematics and physics.

3.2.1 Mathematical Model Method

At the mathematical level, Sarkozy first used nth order polynomials to fit minor nonlinearities, which greatly simplified theoretical calculations. Based on this, he made a software package to include band-limiting effects with Gaussian spectral density, and later, the actual measurements of 12 channels in communication satellites were very consistent with his analysis. In 1981, Eng et al. performed a theoretical analysis on the order and type of PIM, and they used the common tree search algorithm to analyze the problem (Figure 3.1) and found that the reverse tree search algorithm was very efficient. After that, he studied the higher-order PIM products and found that the higher-order PIM products exhibited completely different properties from Gaussian noise in the design of digital channels with low BER (Bit Error Ratio), etc. Boyhan, on the other hand, derived the formula of high-order PIM products from Sunde's basic equation, and its calculation results were very consistent with the measurement results. Abuelma'Atti performed a large-signal analysis of the PIM generated at the corrosion interface. His results show that when certain conditions are met, even if the input signal is large, even-order harmonics and even-order PIM products can theoretically be completely eliminated, while odd-order PIM products will double. As an example, he analyzed the performance of directional coupler under large signals.

At the physical level, foreign researchers have also made a lot of new explorations, and their research mainly follows the two ideas of equivalent circuit method and field analysis method.

3.2.2 Equivalent Circuit Method

The equivalent circuit method refers to the establishment of equivalent circuit models for some specific structures to analyze their nonlinear effects. For example, Vicente et al. studied the PIM interference problem at waveguide connections (Figure 3.2) based on the establishment of an equivalent circuit model and analysis of electron tunneling effects; their results showed that the flange surface cleanliness and mechanical properties have a much higher effect on PIM than the flange surface roughness. Russer et al. did a similar study under the consideration of the effect of skin effects on the results. Zelenchuk et al.

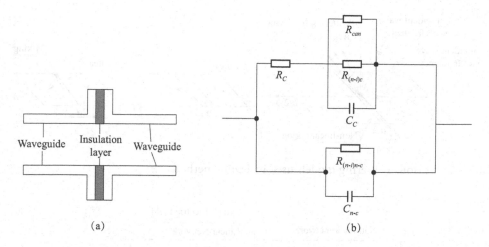

FIGURE 3.2 Study of PIM at waveguide connections. (a) Waveguide interface. (b) Equivalent circuit.

TABLE 3.1 3rd PIM Research Results of Common Coaxial Connectors

Type of Connectors	3rd PIM/dBm	3rd PIM/dBc
SMA connector	−70 to −60	−113 to −103
standard N connector	−85 to −75	−128 to −118
BNC connector	−75 to −70	−119 to 113
silver-plated N connector	−125 to −115	−159 to 167
DIN7~16 connector (≤−125 dBm system residual intermodulation bottom noise)	−140 to −125	−183 to −168

analyzed the PIM characteristics of printed microstrip lines on PCBs using a transmission line model with terminated nonlinear parasitic resistance. The analysis results show that the generation of PIM in microstrip lines is mainly related to nonlinear scattering, attenuation and termination mismatch on the line. When a lossless transmission line is terminated with a matched load, nonlinear scattering is the root cause of PIM. At this point, the 3rd order PIM product can be reduced by selecting a suitable transmission line length. Henrie et al. performed PIM analysis and predictions on various commonly used coaxial connectors and verified them through experiments, as shown in Table 3.1. The number of each type of connector ranges between 1 and 22.

3.2.3 Field Analysis Method

The field analysis method refers to the study of PIM electromagnetic scattering by the field point of view. For example, since 1999, Bolli and Selleri et al. have published several papers on the Time-Domain Physical Optic (TDPO) method. This method can be used to analyze the electromagnetic scattering of PIM products at the interface. Since the number of variables required in their proposed model (Figures 3.3) is small, the method is well suited for embedding in existing CAD software for electromagnetic fields. Later, they also introduced a genetic algorithm to optimize the parameters.

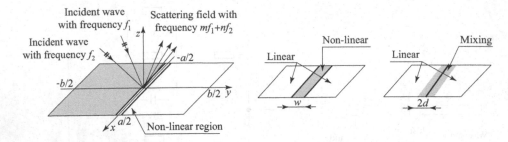

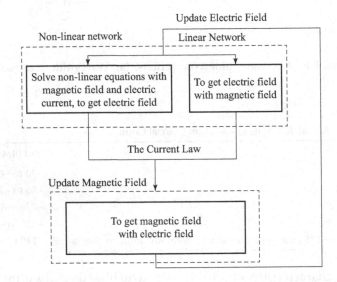

FIGURE 3.3   Various scattering models used in TDPO method.

FIGURE 3.4   Schematic diagram of linear FDTD algorithm flow. (a) EMF advance process. (b) signal pair elimination process.

With the continuous development of computer technology, more and more people are using computer simulation to study. Lojacono et al. used Time-Frequency Representations (TFR) method to simulate the 3rd order PIM products in multichannel transmission systems. Based on the simulation results, they also gave certain design guidance suggestions. Ishibashi et al. simulated current-type PIM interference using the nonlinear Finite-Difference Time-Domain (FDTD) method. In their publication, they assumed that the resistivity is nonlinear and introduced the 3rd order resistivity coefficients in Maxwell's equations, followed by the electromagnetic simulation using the classical FDTD algorithm, and the flow is schematically shown in Figure 3.4. Finally, the open coaxial line is used to verify the method.

3.3 MULTIPHYSICS FIELD COUPLING ANALYSIS OF PASSIVE INTERMODULATION OF MICROWAVE COMPONENTS

3.3.1 Electrical, Thermal and Mechanical Analysis of Microwave Components

The PIM of microwave components generated by contact-induced metal contact J-V nonlinearity is strongly related to the external ambient temperature of the microwave components, the initial pressure at the contact, and the dynamic changes in the temperature and pressure distribution to which the contact is subjected. Therefore, coupled analysis of the electromagnetic, temperature and stress distribution of microwave components and the establishment of a coupled electric-thermal-stress analysis model are the basis for the PIM analysis and evaluation of microwave components.

There is a coupling effect between the electromagnetic field and the thermal field. Due to electromagnetic loss, the propagation of high-power microwaves in microwave components will produce thermal effects Q inside the microwave components, which include resistive loss Q_{rh} and magnetic loss Q_{ml}, namely

$$Q_{rh} = \frac{1}{2}\mathrm{Re}\left(J \cdot E^*\right) \tag{3.1}$$

$$Q_{ml} = \frac{1}{2}\mathrm{Re}\left(j\omega B \cdot H^*\right) \tag{3.2}$$

The resistive loss Q_{rh} can be expressed by the loss tangent of the medium $\tan\delta$ or the imaginary part of the dielectric constant ε'', which is related to the resistivity of the material. The larger the resistivity, the smaller the resistive loss. The magnetic loss Q_{ml} mainly exists in magnetic materials, which is proportional to the imaginary part of the complex permeability μ''.

The heat generated by the electromagnetic loss heat effect is one of the field sources of the temperature field. In addition, the temperature change caused by the external environment is also a major field source of the temperature field. The distribution of the temperature field can be determined by solving the heat transfer equation. However, the effect of the temperature field on the structure is manifested in the expansion or contraction of the object due to the temperature difference, resulting in thermal strain, and their relation is

$$\varepsilon_{inel} = \alpha\left(T - T_{ref}\right) \tag{3.3}$$

where ε_{inel} is the strain generated by the temperature change; α is the coefficient of thermal expansion; T is the temperature on parts; T_{ref} is the strain reference temperature.

The thermal strain is used as the field source of the force field. The distribution of strain ε, displacement u and stress s can be obtained by solving a system of linear elastodynamic equations. These three physical fields are coupled by providing physical field sources to each other.

It should be pointed out that the generation of thermal stresses is related to temperature variation and confinement. Thermal stresses are not generated when the structure undergoes free deformation under temperature change but only when the free deformation is constrained. In addition, if the temperature is not uniformly distributed inside the same object, then even if the object is not constrained by the outside, thermal stress will be generated because the temperature is different everywhere, and each part is not free to expand and contract by the constraint of the neighboring parts at different temperatures. Therefore, the differential equations of these three physical fields (Maxwell's equations or

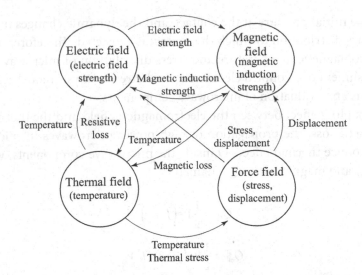

FIGURE 3.5   Schematic diagram of electric-thermal-stress multiphysics coupling relationship.

fluctuation equations, heat transfer equations, and linear elastodynamic equations) can be coupled together by specific physical quantities and relations.

Figure 3.5 illustrates the coupling of the electric-thermal-stress multiphysics field, in which a physical field is indicated in the circle and the basic field variables of the field are listed in the parentheses. The directed line segments indicate the unidirectional action between the physical fields, and the text next to the line segments indicates the physical quantities, e.g., thermal stress indicates that the thermal field is acting on the force field through thermal stress.

1. By solving the fluctuation equations in the electric field form (additional boundary conditions), the electric field distribution in the microwave components can be obtained, with which the displacement, magnetic induction strength, magnetic field strength, free current density and other physical quantities using Maxwell's equations and the intrinsic relationship can be derived.

2. The resistive loss and magnetic loss can be obtained by the formulas (3.1) and (3.2); in the combination of the temperature difference caused by the external environment, the source of the thermal field can be determined, so that the temperature field distribution can be determined by solving the heat conduction equation.

3. The change in temperature causes thermal strain, the thermal stress field source of the force field can be obtained from equation (3.3), and then the distribution of physical quantities such as displacement, strain and stress can be obtained by solving the linear elastodynamic equations.

In addition, the temperature field also affects the material parameters of the model. Some material parameters such as material density, heat capacity and Poisson's ratio are not sensitive to temperature, and they can be regarded as constants when the temperature changes, usually taking their values at room temperature; other physical parameters (including electrical conductivity, thermal conductivity, Young's modulus, and coefficient of thermal expansion) are functions of temperature. Coupling the equation sets of electromagnetic, thermal and force fields together, a coupled analytical model can be obtained with the following equations:

1. Electromagnetic distribution:

$$\begin{cases} \nabla \cdot J + \dfrac{\partial \rho_v}{\partial t} = 0 \\[2mm] \nabla \cdot B = 0 \\[2mm] \nabla \cdot D = \rho_v \\[2mm] \nabla \times E + \dfrac{\partial B}{\partial t} = 0 \\[2mm] \nabla \times H = J + \dfrac{\partial D}{\partial t} \\[2mm] \nabla \times \mu_r^{-1} (\nabla \times E) - k_0^2 \left(\varepsilon_r - \dfrac{j\sigma}{\omega \varepsilon_0} \right) E = 0 \end{cases} \tag{3.4}$$

2. Temperature distribution:

$$\rho C_p \frac{\partial T}{\partial t} + \rho C_p u \cdot \nabla T = \nabla \cdot (k\nabla T) + Q \tag{3.5}$$

3. Mechanics of elastic materials:

$$\begin{cases} s - s_0 = C : (\varepsilon - \varepsilon_0 - \varepsilon_{inel}) \\[2mm] \varepsilon = \dfrac{1}{2} \left((\nabla u)^T + \nabla u \right) \end{cases} \tag{3.6}$$

4. Electromagnetic-thermal coupling relations:

$$\begin{cases} Q = Q_{rh} + Q_{ml} \\[2mm] Q_{rh} = \dfrac{1}{2} \mathrm{Re} \left(J \cdot E^* \right) \\[2mm] Q_{ml} = \dfrac{1}{2} \mathrm{Re} \left(j\omega B \cdot H^* \right) \end{cases} \tag{3.7}$$

5. Thermal-mechanics coupling relation.

$$\varepsilon_{inel} = \alpha \left(T - T_{ref} \right)$$

3.3.2 Electromagnetic-Thermal Coupling Analysis

The microwave thermal mode combines the electromagnetic wave mode with the heat transfer mode. Since microwave energy is absorbed by dielectric materials and converted into heat, the electromagnetic loss of electromagnetic waves can be used as a heat source based on the assumption that the electromagnetic period is shorter than the thermal time scale.

The steady-state simulation is performed using the frequency-domain-steady-state solution mode; the transient simulation is performed using the frequency-domain-transient-state solution mode. In the frequency domain-steady-state solution mode, Maxwell's equations of the heat transfer equations in the steady state are solved in the frequency domain under the assumption that all thermodynamically relevant initial transient variables have vanished. Since there is no transient information, its calculation is much faster than the analysis with the frequency domain-transient solution mode, and the results give the steady-state temperature field distribution. In the frequency domain-transient solution mode, Maxwell's equations of the heat transfer process in the transient state are solved in the frequency domain under the assumption that the material properties used to solve Maxwell's equations remain constant during a single electromagnetic wave oscillation period. The electromagnetic field is recalculated when the material properties change significantly, and this calculation process is determined by the relative tolerance criteria of the time-varying solver.

Electromagnetic waves propagate through the medium of microwave components, which can be set in the wave equation in the form of an electric field to introduce the wave equation. Due to the existence of electromagnetic losses, the propagation of high-power microwaves in microwave components will produce thermal effects inside the microwave components, including electrical and magnetic losses. The electric loss can be calculated based on the complex permittivity, and the finite permittivity is introduced into the complex permittivity as

$$\varepsilon_c = \varepsilon - j\frac{\sigma}{\omega} \tag{3.8}$$

or

$$\varepsilon_c = \varepsilon_0\left(\varepsilon' - j\varepsilon''\right) \tag{3.9}$$

where ε is the dielectric constant of the medium; ε_0 is the vacuum dielectric constant; ε' and ε'' are the real and imaginary parts of the effective dielectric constant ε_r. The electric loss is determined by the imaginary part ε''.

The complex dielectric constant can also be given by the loss tangent as

$$\varepsilon_c = \varepsilon_0\varepsilon'\left(1 - j\tan\delta\right) \tag{3.10}$$

The magnetic loss can be measured from the complex permeability as

$$\mu_c = \mu' - j\mu'' \tag{3.11}$$

Where μ' and μ'' are the real part and imaginary part of permeability, respectively.

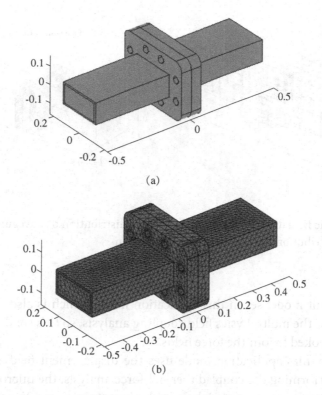

FIGURE 3.6 Rectangular waveguide flange junction. (a) Geometric structure. (b) Meshing.

In devices such as waveguide flange junctions, TNC coaxial connectors and coaxial filters, the magnetic loss can basically be regarded as zero. The metallic part of the microwave component has a large conductivity, $\sigma/(\omega\varepsilon)\gg1$, and the electromagnetic field is mainly concentrated in a thin layer with a thickness equal to the skin depth of the surface, so the electric loss of the microwave component can be introduced by setting the impedance boundary condition.

When setting up the microwave thermal model for the device, the electromagnetic loss Q can be used as the heat source of the thermal field, so that the electromagnetic field can be coupled with the thermal field. In the spaceborne environment, the external environment of the microwave components is vacuum, there is no thermal convection and conduction between the surface and the environment, and the heat exchange between the surface and the environment is carried out in the form of radiated heat. The waveguide flange junction is used as an example for electrothermal coupling simulation, in which the transmitted electromagnetic wave frequency is 0.8 GHz, and the power is 500 W. Figure 3.6 shows the geometric structure and meshing. The calculated results of electric field and temperature distribution of the waveguide flange junction are shown in Figure 3.7.

3.3.3 Thermal-Mechanical Coupling Analysis

The analysis of the coupled thermal and force fields is performed using the solid mechanics application model, which contains the mathematical models and characteristics of stress analysis and general linear and nonlinear solid mechanics. The linear-elastic material

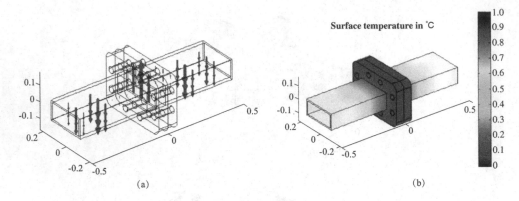

FIGURE 3.7   Electric field distribution and temperature distribution of waveguide flange junction. (a) Electric field distribution. (b) Temperature distribution.

model is the default model set in this application mode, which is also the mathematical model required for the multiphysics field coupling analysis, so this solid mechanics application mode is invoked to join the force fields.

The solid mechanics application mode uses the displacement field as the dependent variable, when performing the coupled thermal-force analysis, the microwave component is regarded as a linear elastic material, set to be anisotropic, and Young's modulus and Poisson's ratio are the key parameters. In the subnode of the linear elastic material, thermal expansion is set to define the internal thermal strain caused by the temperature change, and the strain reference temperature should be consistent with the ambient temperature.

According to the actual situation of microwave components, fixed surfaces are set, and the displacement of these surfaces is zero in all directions. Other boundaries can be set as free boundaries (default force field boundary conditions) if there are no constraints (or loads) in any direction. Boundary loads are added to the microwave components if necessary.

By using unified physical quantities of electromagnetic, temperature and force fields, the electromagnetic field is coupled with the thermal field, the internal thermal strain caused by the temperature change is defined, and the thermal field is coupled with the force field. The mechanical boundary conditions such as fixed surface and boundary load are set according to the actual situation of microwave components, and the stress-strain distribution of the waveguide flange junction is calculated, as shown in Figure 3.8.

3.3.4 Experiment and Analysis of Multiphysical Field Coupling of Microwave Components

In order to improve the accuracy of the multiphysics analysis of microwave components, it is also necessary to correct the multiphysics coupling analysis based on the temperature and stress experimental values of the calibration points on the components. The process is shown in Figure 3.9. First, the geometry of the microwave component is modeled numerically, and material parameters are assigned to establish boundary conditions to carry out multiphysics field coupling analysis for the distribution of temperature and stress on the

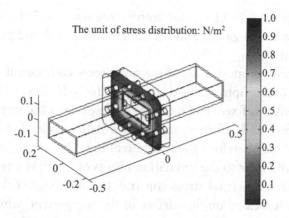

FIGURE 3.8   Stress-strain distribution of waveguide flange junction.

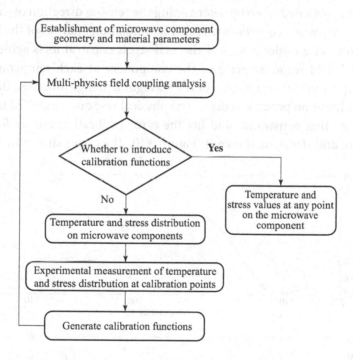

FIGURE 3.9   Multiphysics field experimental calibration analysis process.

microwave component. The temperature and stress distribution of the calibration point are measured experimentally by multiphysics field parameter measurements, the error calibration function is constructed and the analysis algorithm is modified to perform multiphysics field analysis again to obtain more accurate multiphysics field analysis results.

Based on the PIM test system, a multiphysical parameter measurement module was introduced to carry out multiphysical parameter testing of microwave components. The temperature range of the chamber used in this experiment is −40~80°C, the fastest heating rate is about 1.5°C per minute, the fastest cooling rate is about 2°C per minute, and the stable temperature accuracy is 0.1°C. The temperature sensor is a thermocouple, the stress sensor adopts a stress strain gauge, and the data are collected and processed by the data acquisition equipment.

The physical response of most areas of the microwave component surface is small due to the different areas of multiphysics field coupling effects. However, the accuracy and linearity range of the sensor is fixed, and if the sensor (especially the strain gauge) works in a position with relatively small deformation, then it will directly lead to the poor accuracy of the measurement results. Therefore, the selection of the calibration point on the surface of each component should refer to the simulation results of the PIM analysis and evaluation platform under electrical-thermal-stress constraints, and consider the following factors: the calibration point is located on the surface of the component, which is convenient for fixing the sensor and does not affect the PIM characteristics of the component; the calibration point is located in an area with significant deformation, which enables the sensor to operate in the best dynamic range; through simulation analysis, the strain trend of the calibration point is obtained, thereby determining the sensing direction of the strain gauge

For a specific microwave component, a reliable simulation analysis of the PIM and multiphysical field response under electrical-thermal-stress constraints is achieved by fitting the multiphysical field response error of the component at each temperature sampling point. For the experimental verification of the waveguide flange junction, this experiment uses the above calibration process to derive the physical response and PIM under the multiphysics field coupling constraint, and fits the error calibration curves for the component temperature and stress, as shown in Figure 3.10. The curve shown in Figure 3.10a is

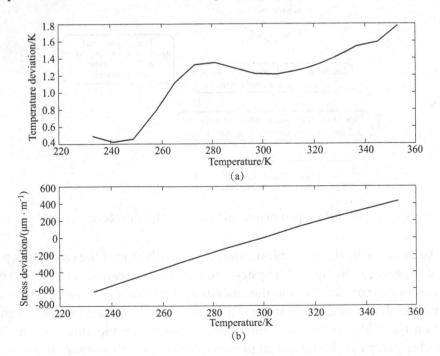

FIGURE 3.10 Error calibration function of temperature and stress of waveguide flange junction. (a) Temperature error calibration function. (b) Stress error calibration function.

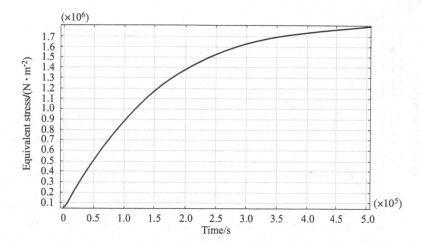

FIGURE 3.11   Temperature variation curve at the highest inner waveguide wall temperature on the junction contact surface of the waveguide junction flange.

the simulated error calibration fitting curve obtained by experimentally measuring the calibration points between 230 and 353 K for the waveguide flange junction. Once the error calibration curve is obtained, the simulation results can be calibrated to obtain the experimentally verified calibration analysis results. After that, multiphysics field response analysis can be achieved at any point on the part surface and at any ambient temperature.

We take the waveguide flange junction as an example; the port input power is 25W, the frequency is 2.4GHz, the propagation mode is TE_{10} mode, the time is $0\sim5\times10^5$s, and the time step is 1000s, and the temperature variation of the highest temperature point of the inside of the waveguide wall on the contact surface of the flange junction obtained from the simulation is shown in Figure 3.11. It can be seen that the temperature rise rate of the waveguide flange junction is the fastest in the first 1.5×10^5s, and the rate decreases gradually thereafter. Since heat conduction is slow, it takes a certain time to reach temperature equilibrium.

The variation of stress with time at the maximum stress on the inner side of the waveguide wall at the flange junction contact surface is shown in Figure 3.12.

The simulation results show that both the temperature field and the stress field show a steady state equilibrium after a certain period of time.

3.4 PASSIVE INTERMODULATION EVALUATION OF MICROWAVE COMPONENTS

3.4.1 Multiscale Electromagnetic Calculation of Microwave Components

Since PIM occurs at the tiny contacts of microwave components, it is necessary to use nonuniform meshing in the numerical analysis of PIM. The FDTD algorithm of nonuniform mesh is improved on the traditional FDTD algorithm, and the mesh is generated according to different rules to make the mesh size of the entire calculation area different. There are also many types of meshing methods, among which the refined mesh technique and

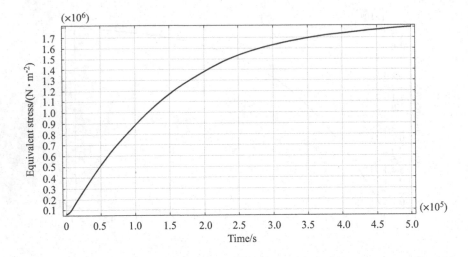

FIGURE 3.12   The stress curve of the maximum point on the inner side of the waveguide wall on the contact surface of the flange.

the variable mesh step method are widely used. Mesh refine technique refers to dividing the area with higher calculation accuracy into fine meshes and dividing other calculation areas into coarse meshes. At present, a variety of methods for information exchange between coarse and fine meshes have been proposed. Kunz et al. proposed a method of submeshing technique twice. The process is that first, the coarse mesh is divided into the entire calculation area for calculation; then, the fine mesh area is calculated, and the field at the cut-off boundary of the coarse and fine mesh is obtained by interpolation in the coarse mesh. The variable mesh step method refers to as follows: in space, a gradual step is taken to adjust the spatial step in combination with the spatial variation of the material structure of the simulated object; in time, the same time step is taken. In terms of the characteristics of the meshing, it can be divided into abrupt meshing and gradient meshing. Abrupt meshing means that the mesh is divided according to the needs of the target structure without rules; gradient mesh means that the mesh is divided according to certain rules, and an expansion factor is added. The number of meshes increases with the power of the expansion factor, and the proportion of adjacent meshes is the same. Obviously, the nonuniform mesh FDTD algorithm is an inevitable choice for simulating the PIM effect. It has high calculation accuracy, saves calculation memory, and the calculation efficiency is much higher than the traditional uniform FDTD algorithm. According to the accuracy of various nonuniform algorithms, this section selects the gradient nonuniform mesh as the core algorithm of the PIM effect numerical simulation. According to the different field values at the boundary of the coarse and fine meshes, the calculation methods of the gradient nonuniform mesh include virtual magnetic field method and integral method.

When the electromagnetic problems are analyzed by nonuniform mesh FDTD algorithm, the division of the mesh size is particularly critical. The division principle is that in the calculation area where the target structure is more complex and the electromagnetic distribution varies greatly, a fine mesh is used; in the calculation area where the target structure is simpler and the electromagnetic distribution does not vary much, a coarse mesh is used. After dividing the whole calculation area into coarse and fine mesh areas, the FDTD algorithm is then used to simulate the target of interest. In the respective calculation areas, calculations can be performed in the differential format of the conventional method; at the boundaries involving the coarse and fine meshes, the tangential components of the electric field can be obtained by interpolation in time and space; the values of each magnetic field on the boundaries can be solved by increasing each time step from the adjacent time and space averages. However, at the coarse and fine mesh boundaries, first-order errors arise due to constant iteration and continuous updating of adjacent mesh nodes. This is because the recursive formulation of the conventional method is based on the central difference approximation solution, and when on a nonuniform mesh, the electric field shifts the two central positions of the magnetic field, i.e., a calculation error occurs at these nodes due to the shift of the electric field components. In the whole calculation area, it will be deviated from the node position at the next time step of the magnetic field value because of the presence of the error. The electric field component is iteratively generated by the four surrounding magnetic field components, which will cause the nodes of the electric field component to shift, thus increasing the calculation error as time increases. In order to reduce the calculation error and improve the accuracy, the magnetic field components can be assumed at the boundary of the coarse and fine meshes. A hypothetical magnetic field component $H'_y(i, j,k)$ is inserted at the boundary of the coarse and fine meshes, and the node of this magnetic field component is located half a mesh node away from the boundary of the coarse and fine meshes, as shown in Figure 3.13.

The assumed magnetic field components $H'_y(i, j, k)$ can be derived from the coarse-mesh magnetic field components and fine-mesh magnetic field components at the boundaries of adjacent coarse and fine meshes, i.e., $H_y(i-1, j, k)$ and $H_y(i, j, k)$.

The magnetic field components $H'_y(i, j, k)$ assumed on the right-hand side of the coarse and fine mesh boundaries are

$$H'_y(i,j,k) = \frac{a-1}{a+1} H_y(i-1,j,k) + \frac{2}{a+1} H_y(i,j,k) \tag{3.12}$$

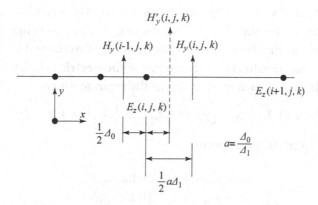

FIGURE 3.13   Virtual magnetic field method.

The magnetic field components $H'_y(i, j, k)$ assumed on the left-hand side of the coarse and fine mesh boundaries are

$$H'_y(i, j, k) = -\frac{a-1}{2a} H_y(i-2, j, k) + \frac{3}{2} H_y(i-1, j, k) \tag{3.13}$$

where a is the ratio of coarse to fine mesh, i.e., the mesh ratio.

As time increases, the field values of the electric field components at the coarse and fine mesh boundaries can be obtained from the magnetic field components adjacent to them; for the electric and magnetic field components that are not on the coarse and fine mesh boundaries, the electric and magnetic field components in the respective calculation areas can be obtained by the FDTD algorithm with a uniform mesh. In summary, the steps of the iterative calculation of the electromagnetic field in the conventional nonuniform FDTD algorithm are as follows.

In the first step, when the time step is $t = n+1/2$, the FDTD difference format of uniform mesh is used to iterate the magnetic field components, and for the hypothetical magnetic field components $H'_y(i, j, k)$ at the boundary of the coarse and fine meshes can be obtained by using interpolation methods.

In the second step, when the time step is $t = n$, the FDTD difference format of uniform mesh is used to iterate the electric field, and the electric field component $E_z(i, j, k)$ at the boundary of the coarse and fine mesh can be obtained by the magnetic field component $H'_y(i, j, k)$ and $H_y(i-1, j, k)$.

In the third step, return to Step 1, and perform iterative calculation of magnetic field.

When meshing the entire calculation area with the coarse and fine meshes, a hypothetical magnetic field is introduced at the boundary of the coarse and fine meshes, which reduces the reflection error caused by the sudden change of the mesh at the boundary of the coarse and fine meshes; however, the calculation is still inaccurate. In different calculation models, the ratio of the coarse and fine mesh size is different, which makes it impossible to determine an optimal ratio. The meshes in the calculation area are divided according to a certain rule, that is, the space step in the calculation area gradually changes with the simulated target, and the time step is the same. In this case, there is only one gradual change factor, so that the divided mesh can satisfy stability requirements and then avoid the introduction of virtual magnetic fields for auxiliary calculations at the discontinuities of coarse and fine meshes. In the case of using gradient meshing, the renewal of the electromagnetic field in the discontinuous mesh can be determined by an integral equation. In the rectangular coordinate system, suppose the electric field E is on the main mesh line, and the coordinates of each mesh point in the main mesh are

$$\{x_i; i = 1, 2, \ldots, N_x\}, \{y_j; j = 1, 2, \ldots, N_y\}, \{z_k; k = 1, 2, \ldots, N_z\} \tag{3.14}$$

The side length between the mesh points is

$$\begin{cases} \Delta x_i = x_{i+1} - x_i; & i = 1, 2, \ldots, N_x - 1 \\ \Delta y_j = y_{j+1} - y_j; & j = 1, 2, \ldots, N_y - 1 \\ \Delta z_k = z_{k+1} - z_k; & k = 1, 2, \ldots, N_z - 1 \end{cases} \tag{3.15}$$

The center coordinates of each mesh are

$$x_{i+\frac{1}{2}} = x_i + \frac{\Delta x_i}{2}, y_{j+\frac{1}{2}} = y_j + \frac{\Delta y_j}{2}, z_{k+\frac{1}{2}} = x_k + \frac{\Delta z_k}{2} \tag{3.16}$$

The side length between each mesh point in the auxiliary mesh is

$$\begin{cases} h_i^x = \dfrac{\Delta x_i + \Delta x_{i-1}}{2}; i = 2,3,\ldots, N_x \\[2mm] h_j^y = \dfrac{\Delta y_j + \Delta y_{j-1}}{2}; j = 2,3,\ldots, N_y \\[2mm] h_k^z = \dfrac{\Delta z_k + \Delta z_{k-1}}{2}; k = 2,3,\ldots, N_z \end{cases} \tag{3.17}$$

The components of electric and magnetic fields are expressed as

$$E_x^n\left(i+\frac{1}{2}, j, k\right) = E_x\left(x_{i+\frac{1}{2}}, y_j, z_k, n\Delta t\right) \tag{3.18}$$

$$H_x^{n+\frac{1}{2}}\left(i, j+\frac{1}{2}, k+\frac{1}{2}\right) = H_x\left(x_i, y_{j+\frac{1}{2}}, z_{k+\frac{1}{2}}, \left(n+\frac{1}{2}\right)\Delta t\right) \tag{3.19}$$

The FDTD difference form of gradually nonuniform meshes is derived from the integral form of Maxwell's equations:

$$\oint_L E \cdot dl = -\frac{\partial}{\partial t}\iint_S B \cdot ds - \iint_S M \cdot ds \tag{3.20}$$

$$\oint_L H \cdot dl = \iint_S J^s \cdot ds + \iint_S \sigma E \cdot ds + \frac{\partial}{\partial t}\iint_S D \cdot ds \tag{3.21}$$

If the metal region in the solved problem does not occupy a large proportion or the solved problem is of moderate scale, then the traditional processing method is competent; otherwise, there is a large room for improving the calculation efficiency of the traditional processing method. In the PIM effect simulation analysis, in order to accurately simulate the geometric and electromagnetic characteristics of the contact surface, the mesh size at the contact surface must be much smaller than $\lambda/10$ (λ is the wavelength); in order to avoid the numerical instability caused by the rapid change of the mesh size, it is necessary to use the gradient mesh to meet certain rules for generation, and the final number of meshes will be huge and may cause numerical instability. Therefore, the simulation efficiency of PIM effect is difficult to meet the actual engineering requirements.

Accordingly, this chapter proposes an adaptive iteration strategy, which uses the metal occupation ratio size to determine the specific iteration method. This strategy ensures the traditional advantages of the algorithm, while significantly improving the calculation efficiency of the time-domain finite-difference algorithm, and also reduces the difficulty of the algorithm design for the subsequent iterative solution of the nonlinear equations. In the traditional time-domain finite-difference algorithm, the iterative equation of the electromagnetic field is as follows:

$$\hat{e}_d\Big|_{i,j,k}^{t+\Delta t} = CAP \cdot \hat{e}_d\Big|_{i,j,k}^{t} + CAN\left(\bar{\tilde{h}}_{d_{rim}}\Bigg|_{\langle i,j,k\rangle_{d_{dex}}+1}^{t+\Delta t/2} - \bar{\tilde{h}}_{d_{m-1}}\Bigg|_{\langle i,j,k\rangle_{d_{prias}}+1}^{t+\Delta/2}\right) \tag{3.22}$$

$$\hat{h}_d\Big|_{i,j,k}^{t+\Delta t/2} = CBP \cdot \hat{h}_d\Big|_{i,j,k}^{t-\Delta t/2} + CBN\left(\bar{e}_{d_{\alpha x}}\Big|_{\langle i,j,k\rangle_{d_{piac}}+1}^{t} - \bar{e}_{d_{pm}}\Big|_{\langle i,j,k\rangle_{d_{net}}+1}^{t}\right) \tag{3.23}$$

where

$$CAP = \left(\frac{\bar{\varepsilon}}{\Delta t} - \frac{\bar{\sigma}}{2}\right)\Bigg/\left(\frac{\bar{\varepsilon}}{\Delta t} + \frac{\bar{\sigma}}{2}\right) \tag{3.24}$$

$$CAN = \left(\frac{l_k}{\tilde{s}_k}\right)\Bigg/\left(\frac{\bar{\varepsilon}}{\Delta t} + \frac{\bar{\sigma}}{2}\right) \tag{3.25}$$

$$CBP = \left(\frac{\bar{\mu}}{\Delta t} - \frac{\bar{\kappa}}{2}\right)\Bigg/\left(\frac{\bar{\mu}}{\Delta t} + \frac{\bar{\kappa}}{2}\right) \tag{3.26}$$

$$CBN = \left(\frac{\tilde{l}_k}{s_k}\right)\Bigg/\left(\frac{\bar{\mu}}{\Delta t} + \frac{\bar{\kappa}}{2}\right) \tag{3.27}$$

In Equations (3.23) to (3.28), $\bar{\varepsilon}$ is the permittivity; $\bar{\sigma}$ is the conductivity; Δt is the time step; l_k is the network length of the marker point k; s_k is the area of the marker point k; $\bar{\mu}$ is the magnetic permeability; $\bar{K}$ is the magnetic field loss. The subscript $d(d = x, y, z)$ denotes a certain coordinate variable, while d_{prior} and d_{next} denote the previous and the next spatial variables, respectively, and satisfy the right-handed circular law $(x{\rightarrow}y{\rightarrow}z{\rightarrow}x)$. For example, when $d = x$, then $d_{prior}{=}z$ and $d_{next}{=}y$. Also

$$\bar{\tilde{h}}_z\Bigg|_{\langle i,j,k\rangle,+1}^{t+\Delta t/2} = \hat{h}_z\Big|_{i,j+1,k}^{t+\Delta t/2} - \hat{h}_z\Big|_{i,j,k}^{t+\Delta t/2} \tag{3.28}$$

$$\bar{\tilde{e}}_y\Bigg|_{\langle i,j,k\rangle_i +1}^{t} = \hat{e}_y\Big|_{i,j,k+1}^{t} - \hat{e}_y\Big|_{i,j,k}^{t} \tag{3.29}$$

For the waveguide structure shown in Figure 3.16, the waveguide wall is filled with metal meshes in the traditional finite difference time domain algorithm. The mesh parameters are CAP=1, CAN=0, CBP=1, CBN=0, and participate in calculations like ordinary meshes. In this case, formulas (3.28) and (3.29) can be simplified as

$$\hat{h}_d\Big|_{i,j,k}^{t+\Delta t/2} = \hat{h}_d\Big|_{i,j,k}^{t-\Delta t/2} \tag{3.30}$$

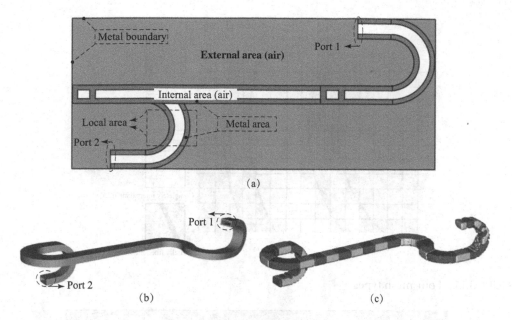

(a)

(b)   (c)

FIGURE 3.14   Waveguide structure in PIM analysis. (a) Electromagnetic modeling of waveguide structures. (b) Geometric schematic of waveguide structures. (c) Electromagnetic field distribution of waveguide structures.

$$\left.\hat{e}_d\right|_{i, j, k}^{t+\Delta t} = \left.\hat{e}_d\right|_{i, j, k}^{t} \tag{3.31}$$

In the conventional time-domain finite-difference algorithm, the initial field values at these meshes are equal to zero. As the iterations proceed, the field values at these meshes are updated by Equations (3.30) and (3.31). It can be found that the field values at these meshes are equal to zero at any moment, although the field values at these meshes are not physically zero-field by setting to zero but by calculation. More critically, as can be seen from Figure 3.14, although the external region of the waveguide is air, the field is confined to the interior of the waveguide due to the presence of the waveguide walls as well as the absorption boundary at the waveguide ports, so that the external field values are also zero, as verified by the numerical results. Therefore, when analyzing the waveguide structure as shown in Figure 3.14, both the waveguide wall and the external space of the waveguide can be removed from the calculation space.

1. Mesh identification.

As shown in Figure 3.15, according to the material distribution in the mesh, the mesh can be divided into four different types in the PIM effect simulation analysis: metal fully filled area, external air fully filled area, internal air fully filled area and partially filled area.

For different mesh types, combined with the independent nonuniform meshing algorithm developed by this research group, the completely filled and partially filled mesh can be distinguished by this mesh property, since the relative side length value

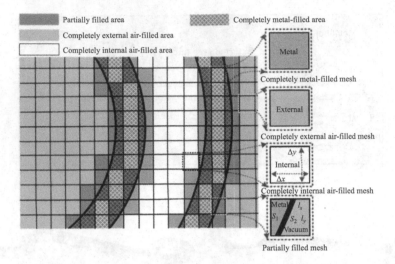

FIGURE 3.15 Four mesh types.

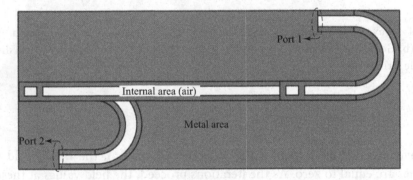

FIGURE 3.16 Schematic diagram of equivalent electromagnetic modeling of waveguide problem.

of the partially filled mesh is less than 1. At the same time, according to the material of the mesh filling, it can be distinguished between the ideal electric conductor full filling and the air fully filled mesh (internal or external). However, for air-filled meshes, the mesh properties of both internal mesh and external mesh are the same, so a new method must be adopted to distinguish them.

2. Equivalent electromagnetic modeling.

Based on the previous analysis, the external area of the waveguide can be replaced with the ideal conductor because the field in the external area does not need to be calculated. As shown in Figure 3.16, the material properties of the external mesh will be changed from completely air-filled to completely metal-filled, and the material properties of the mesh can be judged to distinguish the external mesh from the internal completely air-filled mesh, and the replacement operation will not affect the calculation results.

Finally, after removing all the metal-filled meshes, the final mesh will only contain the internal air-filled mesh and the partially filled mesh. The partially filled mesh uses a conformal meshing technique. Unlike the conventional step approximation technique, the conformal meshing technique allows the waveguide inner wall shape to be accurately described even when the completely filled mesh is removed, thus ensuring the accuracy of the subsequent electromagnetic simulation.

3.4.2 Field Path Combined with Passive Intermodulation Analysis

The nonlinear J-V relationship of the PIM can be regarded as nonlinear resistance. According to the principle of the time-domain finite-difference algorithm, it is known that the time-domain finite-difference algorithm is a simulation of the electromagnetic field (i.e., the field theory is used), and the concept of the lumped element is derived from the circuit, so in order to couple the lumped circuit of the associated circuit concept to the time-domain finite-difference algorithm of the associated field concept, the lumped element modeling method in the Yee mesh is used in this chapter, as shown in Figure 3.17.

Assuming that the electromagnetic wave propagation direction in the microwave component is z-direction, the partial space of the Yee mesh in the y-direction is located inside the metal wall of the microwave component, and the rest is located inside the cavity of the component. In order to analyze the PIM effect due to nonideal metal contact, an equivalent circuit is introduced here (i.e., the lumped elements on the edges of the Yee mesh in Figure 3.17 is replaced with an equivalent circuit), and the nonlinear resistance is used to simulate the nonlinear electromagnetic effect due to electron tunneling at the contact location. The analysis shows that the electromagnetic modeling method can reflect the electromagnetic characteristics of the contact position. In order to ensure the effectiveness of the subsequent simulation algorithm, the electromagnetic modeling method of a typical nonlinear component, a diode, is first studied, and then, on the basis of ensuring that the nonlinear time-domain finite-difference algorithm can accurately simulate this nonlinear component, the results of Chapter 2 (nonlinear J-V relationship) are introduced to further study the general lumped network of electromagnetic modeling and simulation solution

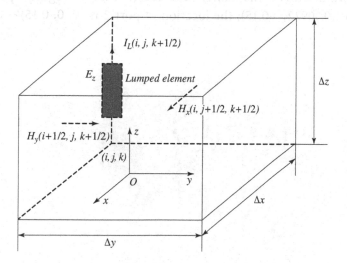

FIGURE 3.17 Modeling method for lumped elements in Yee mesh.

algorithm of the nonlinear time-domain finite difference algorithm, and finally the effectiveness of the algorithm is verified by various ways.

Theoretically, the introduction of nonlinear lumped elements into the time-domain finite-difference algorithm enables the overall full-wave simulation of the PIM effect of spatially complex high-power microwave components. However, in practice, since the level of PIM products is often tens or even hundreds of dB lower than that of linear products, their 3rd order PIM levels are often similar to background noise, and the levels of higher order PIM components are even lower. However, numerical errors are inevitable in the time-domain finite-difference algorithm as a numerical algorithm, so if the nonlinear time-domain finite-difference algorithm is not processed, the PIM products caused by the contact will be overwhelmed by the error noise of the FDTD. Therefore, whether this problem can be effectively solved is the key to determine the success of the numerical analysis of the PIM effect of spatially complex high-power microwave components using nonlinear time-domain finite-difference algorithms.

According to the location and characteristics of the PIM, it is known that the PIM effect is not dominant in the whole microwave component because it occurs only at the nonideal metal-metal contact, and these locations occupy only a very small part of the whole microwave component, which also makes the PIM level very low. Based on this physical mechanism, it can be assumed that the response of the whole microwave component is still dominated by linearity, and the PIM effect is regarded as a perturbation phenomenon, and then the idea of perturbation method in computational electromagnetics is used for the analysis of nonlinear time-domain finite difference algorithm of PIM effect, which will solve this problem to a large extent.

3.4.3 Analysis Example

Next, we take the waveguide flange shown in Figure 3.18 as an example for PIM analysis.

The parameters of the junction of the waveguide flanges are as follows.

Modeling parameters of the flange: the location of waveguide port 1 is (0, 0, −0.15) ~ (0.05461, 0.10922, −0.15); the location of port 2 is (0, 0, 0.15) ~ (0.054 61, 0.109 22, 0.15).

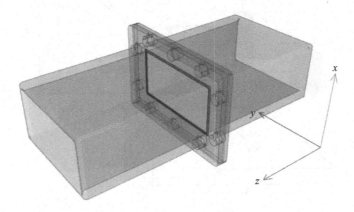

FIGURE 3.18   Waveguide flange.

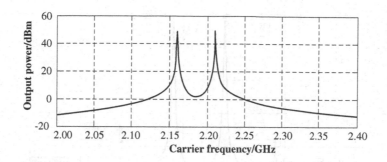

FIGURE 3.19   EM simulation results of rectangular waveguide without contact gap.

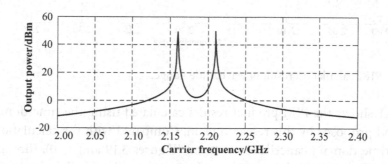

FIGURE 3.20   EM simulation results of rectangular waveguide with contact gap.

Excitation signal: dual carriers with carrier frequencies of 2.16 and 2.21 GHz and total power of 180 W.

Meshing: overall mesh size is 2.7 mm.

Nonuniform meshing: the direction range is −7∼7 mm; the meshing size is 1.73 mm.

Figure 3.19 shows the electromagnetic simulation results of the rectangular waveguide without contact gap. As can be seen from the figure, since the flange connection is removed and there is no micro-contact, no PIM component will appear in the output signal. In the figure, the frequency values of the two main peaks are 2.16 and 2.21 GHz, which are the carrier frequencies of the dual carriers; the output power P_{out} is 49.5 dBm for both, and the numerical simulation is consistent with the theoretical analysis, which verifies the accuracy of the electromagnetic simulation.

Figure 3.20 shows the simulation results of the electromagnetic response of a rectangular waveguide with contact gap. Theoretically, since there is a flange connection, the PIM effect caused by nonideal contact will appear in the waveguide, which is manifested by the appearance of PIM components in the spectrum. However, as can be seen from the figure, the PIM simulation without any treatment does not show any PIM component, which is almost identical to the situation without contact gap as represented in Figure 3.19. Obviously, this is consistent with the previous theoretical analysis. Since there are errors in the numerical calculation and the PIM level is very low, PIM effect cannot be accurately estimated without noise reduction.

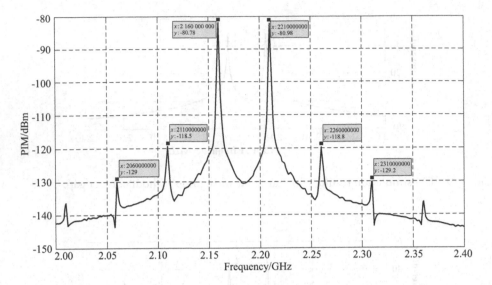

FIGURE 3.21   PIM calculation result of waveguide flange.

Figure 3.21 shows the PIM product results calculated using the time-domain cancellation method proposed by this research team. Compared with the calculation results of the unused time domain cancellation method (Figures 3.19 and 3.20), through the time domain cancellation method, the waveguide flange can be clearly seen in the spectral domain under the excitation of 180W input power. The nonlinear effect is obvious to the PIM components of each order. The third-order PIM is −118.5 dBm, and the experimental result is −122 dBm. The simulation results are in good agreement with the experimental results, which proves the effectiveness of the time-domain cancellation method and further validates the accuracy and effectiveness of the PIM effect prediction method for complex structural microwave components under electrical-thermal-stress constraints developed by this team.

3.5 PASSIVE INTERMODULATION ANALYSIS OF GRID ANTENNA

Satellite communication technology began in the 1960s and flourished in the 1970s and 1980s; in the 1990s, the application of personal mobile communication gave new impetus to the development of satellite communication. Antenna technology is the key technology of satellite communication. With the development of space technology, more stringent requirements are put forward for the spaceborne antenna – narrower beam, higher gain and longer communication distance also make the size of antenna reflective surface and the transmitting power larger and larger. Under the limitation of transmitting volume and transmitting quality, deployable grid reflector antenna is widely used in space application because of its light weight, small size in the retracted state, high deployed-to-stowed ratio and wide frequency range. When the transceiver antenna (or only transmitting antenna) is located near the receiving antenna, the PIM generated by the antenna will have a serious impact on system performance. Therefore, the analysis and evaluation of the grid antenna PIM at the beginning of the antenna design are crucial to the design of the grid antenna.

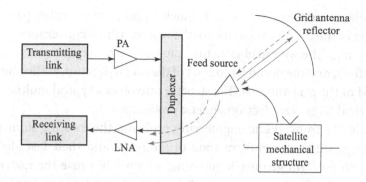

FIGURE 3.22 Block diagram of a satellite-based communication system with a common transceiver for mesh reflector antennas.

Figure 3.22 shows a block diagram of the mesh reflector antenna transceiver-shared satellite-based communication system.

However, the microscopic contact mechanism of metal grids is very complicated, and the theoretical approach of PIM modeling of grid-reflecting antennas is still in its initial stage. In general, in 2002, Pelosi's group proposed that PIM scattering is caused by nonlinear metallic contacts. They also predicted PIM by applying the time-domain physical optics (TDPO) method to nonlinear space. However, the method is not accurate in calculating complex microwave components because the incident wave is approximately plane wave and the TDPO method has many serious cumulative errors. To solve this problem, Ying Liu and Yuru Mao et al. proposed a PIM collaborative calculation method on solid reflector antennas composed of metal plates and gaps based on frequency domain simulation and nonlinear circuits. In 2013, Jie Jiang and Tuanjie Li et al. studied the PIM problem of wire mesh and combined the GW model with Hertz contact theory in the analysis of wire mesh contact nonlinearity to establish the relationship between the contact load and the actual contact area on the rough surface of metal wire. In 2018, Dongwei Wu and Yongjun Xie et al. proposed a PIM prediction method for mesh reflecting surface antennas based on the actual woven mesh reflector antenna, and analyzed the law of influence of metal mesh tension, surface roughness, temperature and other factors for the PIM of mesh transmitting surface antennas.

3.5.1 Electromagnetic Multiscale Equivalent Analysis Method of Grid Antenna

Multiscale electrically large engineering problem refers to the macroscopic scale of the electromagnetic engineering problem reaching tens or even hundreds of electrical wavelengths, but there are microstructures on the scale of a few tenths of wavelengths; these microstructures are small in scale, but because of the huge number and special material (or structure), its influence on the macroscopic properties of the electromagnetic problem is not negligible.

For typical loosely connected structures (such as grid antennas), the grid scale is below a few tenths of a wavelength, and the electrical properties need to be solved in the range of tens or even hundreds of wavelength scales. The meshing density of the microstructure in the multiscale electromagnetic structure is much larger than the other parts, resulting in the deterioration of the condition number of the matrix, which converges very slowly in the solution or even not. The wire grid structure causes electromagnetic wave leakage due to the bypassing effect, and the nonideal contact of the metal junction at the nodes of the wire grid causes PIM of the grid antenna. The structure involves a typical multiscale, nonlinear PIM with electrical large-size electromagnetic problems.

For multiscale electromagnetic engineering problem, the existing numerical calculation methods (e.g., finite elements, method of moments and their fast algorithms, etc.) cannot be used to perform geometric meshing, which will cause the matrix solution to fail to converge, and even the geometric subdivision cannot be completed. However, the macroscopic effects of microstructures must be analyzed and evaluated by numerical calculations due to its number of scales, materials, structures, etc. The equivalence method is an effective means to solve this problem. The equivalence method refers to the equivalence of the effect of the microstructure on the electromagnetic behavior in the macrostructure in the form of boundary conditions and adding the macrostructure for solution. This equivalence method divides the problem into two stages: in stage 1, the electromagnetic response of the microstructure is studied and equated as boundary conditions; in stage 2, the macroscopic electromagnetic problem that does not contain a specific microstructure but contains its equivalent boundary conditions is solved. The equivalence method of the multiscale problem turns difficulty into ease, the computational difficulties caused by the numerical discretization in the direct solution method are solved, and the electromagnetic effects of the microstructure are fully considered. Multiscale electrical large problems of periodic microstructures with loose contacts (e.g., screen reflective surfaces, frequency selective surfaces, radome high frequency band characterization, etc.) can be numerically analyzed by the equivalent analysis method.

Rahmat-Samii et al. in USA applied the Floquet method to study the characteristics of periodic reflective surface antenna (grid antenna) without considering the electrical contact model (i.e., ideal lap) and mainly studied the influence of the size and shape of the periodic structure unit on the reflective characteristics. For the effect of lap connection, two extreme cases of ideal connection and complete disconnection are evaluated as the limits. By the multiscale equivalent analysis method, Yuru Mao and Yongjun Xie et al. performed electromagnetic modeling of loosely connected periodic structures, introduced RLC boundary conditions for the equivalent electrical contact model of metal laps at metal junctions, and used the Floquet method to solve the electrical performance parameters of periodic planar structures. Figure 3.23 shows the calculation process of this multiscale equivalence method for grid antennas.

3.5.1.1 Grid Antenna Wire Contact Modeling Analysis

In fact, the metal-to-metal contact is not a completely smooth surface contact but a surface contact with a certain roughness on the contact surface. The roughness of the metal surface refers to the micro-geometric shape characteristics of some tiny spacing and peaks and valleys that are distributed on the metal processing surface together through machining. In this way, the metal-metal contact is actually a contact between the peaks and valleys

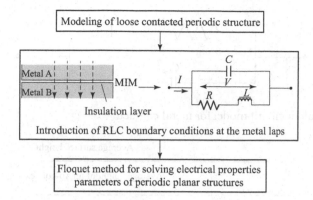

FIGURE 3.23 The calculation process of the grid antenna multiscale equivalent method.

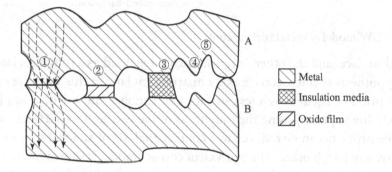

FIGURE 3.24 Rough contact surface section.

on the metal surface, which can also be understood as a combination of a number of contact situations between convex part and concave part. The contact surface is exposed to air, and an oxide layer is formed on the surface or contaminants are adsorbed, forming a Metal Insulator Metal (MIM) structure with a rough contact section as shown in Figure 3.24. There are five contact states in Figure 3.24: ① metal contact; ② metal oxide film sandwiched between contact surfaces; ③ insulating medium sandwiched between contact surfaces; ④ tiny air gap; ⑤ larger air gap. ① and ② form the main channels of current, forming contraction resistance and contact resistance; The oxide in ② conducts relies on the tunneling effect and metal bridge through the film to conduct electricity, which is a semiconductor contact conductive and nonlinear; ③ does not conduct electricity, current passes around to the metal contact; in the air gap (④ and ⑤), current flows around the gap. When the current encounters the impedance Z, a gap voltage is generated, and the gap voltage is potential and may activate any nonlinear effects, resulting in the PIM effect.

The equivalent circuit model for a metal contact is shown in Figure 3.25, with capacitance C_c in the actual contact area, C_{n-c} in the noncontact area and R_c as the contact resistance.

After the metal contact is subjected to external loading, the micro-contact area is approximated as a surface contact. For the rough surfaces contact, the electrical properties directly acting on the contact surface will be studied. When the GW model is used to study the contact state of two rough surfaces, one rough surface is regarded as an ideal

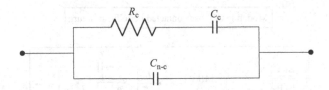

FIGURE 3.25   Equivalent circuit model for metal contact.

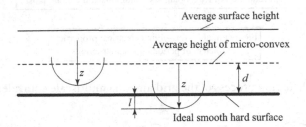

FIGURE 3.26   GW model of metal surface contact.

smooth hard surface, and the other rough surface is regarded as a rough surface conforming to the hypothesis of microconvex body distribution law. In this way, the introduction of the assumption of equal roughness surface can simplify the calculation of the actual contact area. After superimposing the roughness of the two contact surfaces to an equivalent plane, the problem can be reduced to a contact analysis of an ideal smooth hard plane with an equivalent rough plane. The equivalent contact model is shown in Figure 3.26. The micro-convex body of the rough surface is partially involved in the contact and pressed into the smooth and hard plane, generating contact area and bearing contact load; some do not participate in contact and do not produce contact area.

In Figure 3.27, l represents the radial deformation of the micro-convex bodies when the convex peaks of the micro-convex body is pressed onto the hard surface; d represents the distance between the reference line of the average height of all micro-convex bodies on the rough surface and the ideal smooth hard surface; z represents the height of the microconvex body. In addition, A_n represents the nominal contact area; A_{real} represents the actual contact area; the contact surface roughness parameter is $\beta = \eta\sigma r$, η represents the microconvex distribution density, σ represents the microconvex height standard deviation; r represents the microconvex radius; assuming all micro-convex bodies The heights of the convex bodies are all equal, and the distribution probability density function $\varphi(z)$ obeys the Gaussian distribution.

The calculation formula of the dimensionless actual contact area $A^* = A_{real}/A_n$ is as follows:

$$A^* = \pi\beta l_c^* \left(\int_{d^*}^{d^* + l_i^*} I^1 + 0.93 \int_{d^* + l_i^*}^{d^* + 6l_i^*} I^{1.136} + 0.94 \int_{d^* + 6l_c^*}^{d^* + 110l_c^*} I^{1.146} + 2 \int_{d^* + 110l_c^*}^{+\infty} I^1 \right) \tag{3.32}$$

where the expression of I^α is

$$I^\alpha = \left(\frac{z^* - d^*}{l_c^*} \right)^\alpha \phi^*(z^*) dz^* \tag{3.33}$$

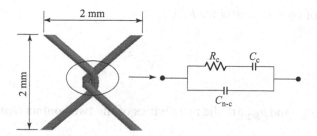

FIGURE 3.27 Electromagnetic model of metal mesh periodic structure with MIM structure of electrical contact model.

where the average height of dimensionless microconvex body $d^*=d/\sigma$; the expression of probability density function $\varphi^*(z^*)$ of dimensionless microconvex body height z^* is

$$\phi^*\left(z^*\right)=\frac{1}{\sqrt{2\pi}}\frac{\sigma}{\sigma_s}\exp\left(-0.5\left(\frac{\sigma}{\sigma_s}\right)^2\left(z^*\right)^2\right)$$

$$\frac{\sigma}{\sigma_s}=\sqrt{1-\frac{3.717\times10^{-4}}{\beta^2}}\qquad(3.34)$$

where σ_s is the standard deviation of equivalent surface height of microconvex body.

The dimensionless critical deformation is $l_c^*=l_c/\sigma$, and the expression of l_c is

$$l_c=\left(\frac{\pi KH}{2E}\right)^2 r\qquad(3.35)$$

where $K=0.454+0.41\nu$, ν is Poisson's ratio of the material, H is the material hardness and E is the Hertzian modulus of elasticity.

After a contact load is applied to the contact surface, the micro-convex bodies will enter into contact, and the contact load determines the number of micro-convex bodies N_c involved in the contact, which is expressed as

$$N_c=\eta A_n\int_{d^*}^{+\infty}\phi^*\left(z^*\right)dz^*\qquad(3.36)$$

Therefore, the expression for the noncontact capacitance C_{n-c} in Figure 3.27 is

$$C_{n-c}=\varepsilon_0 A_n\left(1-A^*\right)d\qquad(3.37)$$

The expression for the contact capacitance C_c is

$$C_c=\frac{\varepsilon_r\varepsilon_0 A_n A^*}{s}\qquad(3.38)$$

where ε_r is the relative permittivity of the insulating layer; s is the thickness of the insulating layer.

The expression of contact resistance R_c is

$$R_c \approx \frac{\rho_{M1} + \rho_{M2}}{4N_c\bar{R}} \tag{3.39}$$

where $\bar{R} \approx \sqrt{\dfrac{A_n A^*}{\pi N_c}}$; ρ_{M1} and ρ_{M2} are the resistivities of the two contact materials.

3.5.1.2 Electromagnetic Analysis of Periodic Woven Structure of Grid Antenna

Due to the periodicity of the woven structured metal mesh, solving the electromagnetic waves on the metal mesh using the Floquet mode expansion implemented based on the finite element method is an effective solution method. The periodic finite element method allows the solution of an infinitely large plane by analyzing only periodic cells. First, a periodic cell model constrained by periodic boundary conditions is built, and the periodic finite element method is used to obtain the reflection coefficients in the infinite plane. The d_x and d_y in the following equations are periodic parameters. For a periodic structure, the electromagnetic field satisfies the periodicity condition of Floquet's theorem. Then, the scattered field E_S and the transmitted field E_T can be written as

$$E_S = \sum_{m=1}^{2} \sum_{p=-\infty}^{+\infty} \sum_{q=-\infty}^{+\infty} R_{mpq}\psi_{mpq} \tag{3.40}$$

$$E_T = \sum_{m=1}^{2} \sum_{p=-\infty}^{+\infty} \sum_{q=-\infty}^{+\infty} B_{mpq}\psi_{mpq} \tag{3.41}$$

If the far field of the scattered field consists of TE and TM modes only, then $p = 0$ and $q = 0$; otherwise, there may be higher-order mode, $p = 0$ and $q = 0$ correspond to the main mode and $p \neq 0$ and $q \neq 0$ correspond to the higher-order mode. The higher-order mode are a fading mode and the field decreases rapidly with increasing distance. The subscript m is 1 and 2 for TE mode and TM mode, respectively. R_{mpq} and B_{mpq} denote the reflection and transmission coefficients of the periodic structure, respectively.

The obtained TE and TM vector mode function ψ_{mpq} has the following form:

TE module:

$$\psi_{1pq} = \frac{1}{\sqrt{d_x d_y}} \left(\frac{v_{pq}}{t_{pq}} \hat{x} - \frac{u_{pq}}{t_{pq}} \hat{y} \right) \varphi_{pq} \tag{3.42}$$

TE module:

$$\psi_{2pq} = \frac{1}{\sqrt{d_x d_y}} \left(\frac{v_{pq}}{t_{pq}} \hat{x} + \frac{u_{pq}}{t_{pq}} \hat{y} \right) \varphi_{pq} \tag{3.43}$$

where

$$\varphi_{pq} = e^{-i\left(u_{m\tau}x + v_{pq}y + \gamma_{pq}z\right)} \tag{3.44}$$

When the electromagnetic wave is incident on the metal grid plane, with the propagation constant K and the incident direction (θ, φ), then

$$u_{pq} = k\sin\theta\cos\phi + 2\pi p / d_x \tag{3.45}$$

$$v_{pq} = k\sin\theta\cos\phi + 2\pi q / d_y \tag{3.46}$$

$$\gamma_{pq} = \begin{cases} \left(k^2 - t_{pq}^2\right)^{1/2}, & k^2 > t_{pq}^2 \\ -i\left|\left(k^2 - t_{pq}^2\right)^{1/2}\right|, & k^2 < t_{pq}^2 \end{cases} \tag{3.47}$$

where

$$t_{pq}^2 = u_{pq}^2 + v_{pq}^2, \quad p, q = 0, \pm1, \pm2, \ldots, \pm\infty \tag{3.48}$$

The electrical properties of a metal mesh depend to a large extent on the contact of these wire contact points. Any factor that impedes the flow of current through these junctions will result in a deterioration of the reflective properties. Therefore, the effect of wire cross-contact should be considered in the modeling of the metal grid surface. To properly model the contact state, the set boundary conditions are applied on the contact area of the contact points as shown in Figure 3.27 (the set boundary condition equivalent circuit is shown in Figure 3.25), thus achieving an accurate solution of the electromagnetic field on the surface of the metal mesh.

3.5.1.3 Simulation Example

For example, using the electromagnetic multiscale equivalent method to analyze a positive-fed parabolic reflector antenna, in which the parabolic diameter is 0.6 m, the focal length is 0.4 m. The pyramid horn is located at the focal point of the reflector, the feed horn VSWR is 1.2, the operating frequency is 8 GHz, and the polarization mode is line polarization. According to the formula in the first part of section 3.5.1, the lumped parameters of the equivalent circuit are calculated as $R_c \approx 50\,\Omega$, $C_{n-c} \approx 56$ pF and $C_c = 25$ pF.

The contact state at the contact junction of the woven structure shown in Figure 3.27 is characterized by three states, which are ideal contact (perfect-computed), no contact (broken-computed) and lumped boundary loading (lumped-computed). The reflection loss of the metal mesh can be derived using the periodic finite element method, as shown in Figure 3.28.

Figure 3.29 gives the far-field radiation direction of the reflective surface antenna with different contact models. The solid reflecting surface is made of the same material as the grid antenna, and the metal grid is woven by copper wire. The solid line indicates the radiation direction of the solid reflecting surface antenna, and the dashed line and the solid line with circle indicate the perfect connection and the lumped connection at the wire contact junction, respectively.

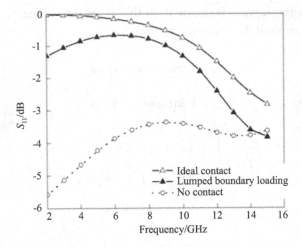

FIGURE 3.28   Metal grid reflection coefficient in different contact states.

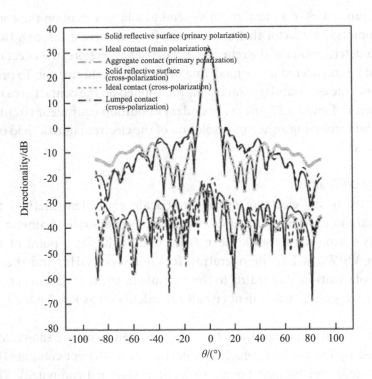

FIGURE 3.29   Far-field radiation direction diagram of reflecting surface antenna with different contact models (8 GHz).

Table 3.2 lists the maximum directionality of these reflective surface antennas. As can be seen from the table, the maximum directionality of the lumped connected reflector antenna is reduced by 0.9 and 0.6 dB at 8 GHz compared with the solid reflector and the ideal contact reflector, respectively. In the actual application of the mesh reflector antenna, the electrical performance characteristics of the metal grid largely depend on the contact

TABLE 3.2   Maximum Directionality of Reflective Surface Antenna

Reflector Antenna	Maximum Directionality /dB
Solid reflective surface	31.7
Ideal contact reflective surface	31.4
Lumped equivalent contact model reflecting surface	30.8

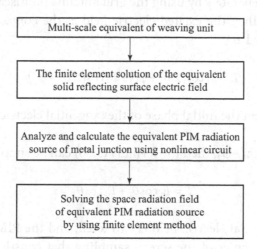

FIGURE 3.30   Block diagram of PIM cosimulation method for grid antenna.

state at the contact junction of these wires. Establishing a proper mesh weaving cell model helps to realize the accurate analysis of the performance of mesh reflecting surface antenna, and the accurate electrical performance analysis is the basis of PIM analysis.

3.5.2 Cosimulation Method for Passive Intermodulation of Grid Antennas

The PIM of the grid antenna is due to the metal contact current-voltage nonlinearity caused by wire contact to generate PIM current, and the PIM current element is then radiated into the receiving channel to form interference. In order to simulate this process and accurately evaluate the PIM performance of the grid antenna, the grid antenna PIM cosimulation method is mainly used, and its flow block diagram is shown in Figure 3.30. In the grid antenna PIM cosimulation, the multiscale equivalence method is used to analyze the reflectivity and other parameters of the actual woven grid antenna; the solid reflective surface of the same reflectivity with lossy material is used to equate the mesh reflective surface, and the finite element method is used to solve the surface electric field of the reflective surface; then, the nonlinear circuit is used to analyze and calculate the equivalent PIM radiation source of the metal junction; finally, by assuming that the metal junction is equivalent to the electric dipole, the total radiated PIM power can be simulated by frequency domain FEM.

The general form of the nonlinear circuit model at the wire contact junction of the grid antenna is

$$H = f(E) = C_0 + C_1 E + C_2 E^2 + C_3 E^3 + \cdots + C_n E^n + \cdots \qquad (3.49)$$

where E represents the electric field strength on the surface of the grid antenna; H represents the magnetic field strength on the surface of the grid antenna; C_n represents a coefficient that depends on the nonlinear characteristics of the metal junction.

Assume that two high-power electromagnetic waves with frequencies f_1 and f_2 are incident on the grid antenna, and the tangential electric field amplitude on the surface of the grid antenna is solved separately by using the grid antenna multiscale equivalent method, denoted as E_1 and E_2. Then the excited electric field at the contact junction of the grid antenna can be expressed as

$$E = E_1 \cos\left(2\pi f_1 + \varphi_1\right) + E_2 \cos\left(2\pi f_2 + \varphi_2\right) \tag{3.50}$$

where φ_1 and φ_2 represent the initial phase of the tangential electric field on the surface of the grid antenna.

Let $\theta_1 = 2\pi f_1 + \varphi_1$, $\theta_2 = 2\pi f_2 + \varphi_2$, then the formula (3.50) can be simplified as

$$E = E_1 \cos\theta_1 + E_2 \cos\theta_2 \tag{3.51}$$

First, the surface tangential electric field is extracted, and the PIM equivalent radiation source is placed at the center of the screen sampling that constitutes the reflector. The direction of the surface tangential electric field E of the reflecting surface is shown in Figure 3.31. The magnetic field H is in the same plane as the electric field E and is perpendicular to the direction of the electric field E. In addition, the number and coordinates of the PIM equivalent radiation source depend on the diameter of the reflecting

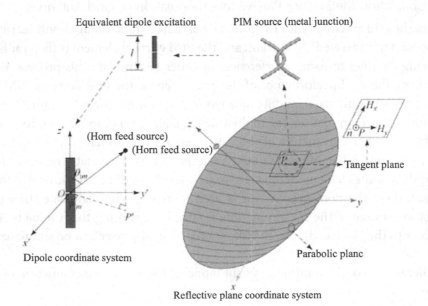

FIGURE 3.31   PIM source equivalence diagram.

surface and the size of the wire mesh sampling area. The coordinate system of reflecting surface (x, y, z) and the coordinate system of dipole (x', y', z') are defined in Figure 3.31. The x-polarization component and the y-polarization component of the PIM of the mesh reflecting surface antenna are simulated separately. Then, the synthetic value of the two PIM components is used as the average value of the PIM. Therefore, a parabolic curve is drawn along the x-axis (or y-axis) over the sampling center of the reflecting surface to simulate and extract the surface tangential electric field on it. Take the PIM radiation source P on the reflective surface shown in Figure 3.31 as an example; its coordinates in the reflective surface coordinate system are defined as $P(x_0, y_0, z_0)$. The projection of the reflecting surface in the xOy plane of the reflecting surface coordinate system is divided into quadrilateral partitions, and the horizontal coordinates (x_0, y_0) of the center position of the quadrilateral partition can be obtained, and the longitudinal coordinates (z_0) can be obtained by substituting the horizontal coordinates (x_0, y_0) into the reflector surface equations. Then, the coordinates of the equivalent PIM radiation source of the quadrilateral partition are (x_0, y_0, z_0).

The parabolic equation passing through $P (x_0, y_0, z_0)$ is

$$z = y^2 / 4F_c, \ y \in \left(-\sqrt{\frac{D^2}{4} - x_0^2}, \sqrt{\frac{D^2}{4} - x_0^2} \right) \tag{3.52}$$

where D represents the diameter of reflector; F_c represents the focal length of reflector
Then, the parabola passing through point P is

$$z - z_0 = y_0 (y - y_0) / (2F_c) \tag{3.53}$$

Therefore, the unit vector of the electric field at point P is

$$\hat{E} = \frac{(0, 1, y_0 / (2F_c))}{\sqrt{1 + (y_0 / (2F_c))^2}} \tag{3.54}$$

The unit normal vector is

$$\hat{n} = \frac{\left(\dfrac{\partial f}{\partial x}, \dfrac{\partial f}{\partial y}, \dfrac{\partial f}{\partial z} \right)}{\sqrt{\left(\dfrac{\partial f}{\partial x} \right)^2 + \left(\dfrac{\partial f}{\partial y} \right)^2 + \left(\dfrac{\partial f}{\partial z} \right)^2}} \Bigg|_{(x_0, y_0, z_0)} = \frac{(x_0, y_0, -2F_c)}{\sqrt{x^2 + y^2 + 4F_c^2}} \tag{3.55}$$

where $f(x, y, z) = x^2 + y^2 - 4F_c z$. Thus, the magnetic field unit vector at point P is $\hat{H} = \left(\hat{n} \times \hat{E} / |\hat{E}|^2 \right)$. The tangent plane shown in Figure 3.33 consists of electric field E and magnetic field H. The dipole is placed along the electric field of the tangent plane (corresponding to the z-axis of the dipole coordinate system).

Substituting equation (3.51) into equation (3.49), we have

$$H = \left(\frac{3}{4} C_3 E_1^2 E_2 + \frac{5}{4} C_5 E_1^4 E_2 + \frac{15}{8} C_5 E_1^2 E_2^3 + \frac{105}{64} C_7 E_1^6 E_2 + \cdots \right) \times \cos(2\theta_1 - \theta_2)$$

$$+ \left(\frac{5}{8} C_5 E_1^3 E_2^2 + \frac{105}{64} C_7 E_1^5 E_2^2 + \frac{35}{16} C_7 E_1^3 E_2^4 + \cdots \right) \times \cos(3\theta_1 - 2\theta_2) + \cdots \tag{3.56}$$

$$J_S = \hat{n} \times H_j, \quad j = 3,5,7,9,\ldots \tag{3.57}$$

$$I_A = \int_l J_S \cdot (\hat{n} \times dl) \tag{3.58}$$

$$\theta_j = m\theta_1 + n\theta_2 \tag{3.59}$$

$$j = |m| + |n| \tag{3.60}$$

Equation (3.60) gives the frequency combination of the PIM product θ_j, where m and n are integers; j is the order of PIM; H_j is the magnetic field component of the jth order PIM (PIM frequency); J_s is the PIM surface current density; $\hat{n}$ is the unit normal vector at the screen sampling center on the reflecting surface; I_A is the nonlinear circuit of the jth order PIM as the excitation current of the dipole to solve for the PIM radiation field.

Based on the radiation angle θ_{im} of dipole i, the coordinates of the horn feed source in the reflecting surface coordinate system are assumed to be (x_m, y_{im}, z_{im}), and the horn feed source is divided into M_i parts with the radiation angle interval of dipole i, which are numbered as m ($m = 1, 2,\ldots, M_i$). The coordinates of the horn port in the dipole coordinate system can be obtained by coordinate transformation. The radiation angles θ_{im} and φ_{im} of the dipole are shown in Figure 3.33. The coordinate transformation relation between the reflecting surface coordinate system and the dipole coordinate system is

$$\begin{bmatrix} x' \\ y' \\ z' \end{bmatrix} = \begin{bmatrix} 1 & 0 & 0 \\ 0 & \cos\alpha & \sin\alpha \\ 0 & -\sin\alpha & \cos\alpha \end{bmatrix} \begin{bmatrix} x - x_0 \\ z - z_0 \\ -(y - y_0) \end{bmatrix} \tag{3.61}$$

where $\alpha = a\,tan_0/(2F_c)$, and F_c is the focal length of the reflector. Therefore, the coordinates of the horn feed source in the dipole coordinate system (x'_m, y'_{im}, z'_{im}) can be derived from equation (3.61), and the radiation angles θ_{im} and φ_{im} of dipole i are

$$\theta_{im} = a\tan\left(\frac{\sqrt{x_{im}'^2 + y_{im}'^2}}{z_{im}'} \right), \quad \varphi_{im} = a\tan\left(\frac{y_{im}'}{x_{im}'} \right) \tag{3.62}$$

The modified radiation field formula is

$$\bar{E}_{im} = j\frac{60\pi I_A l}{\sqrt{2}\lambda r_{im}} \sin\theta_{im} e^{-j(k_m + \varphi_i)} \tag{3.63}$$

where $\bar{E}_{im}, r_{im}$ are the equivalent radiation electric field and radiation distance corresponding to the i-th dipole at radiation angle θ_{im}.

Substituting Equation 3.64 into the radiation power formula of plane wave, the total PIM power received by horn feed source from mesh reflector can be calculated, i.e,

$$P = \sum_{i=1}^{N} P_i = \sum_{i=1}^{N} \sum_{m=1}^{M_i} \left| \bar{E}_{im} \right|^2 S_{im} / (2\eta_0)$$ (3.64)

where P represents the sum of the radiated PIM power of the equivalent dipoles of all contact junctions of the grid antenna; M_i and S_{im} represent the number of face elements on the receiving aperture of the feed of the i-th dipole and the effective receiving area of the m-th face element, respectively; η_0 represents the wave impedance in free space; and N represents the number of dipoles.

The total PIM power of the circularly polarized antenna is the sum of the two linear PIM components (horizontally polarized and vertically polarized) with a phase shift $\frac{\pi}{2}$. In addition, the receiving efficiency of the real horn needs to be considered. In Figure 3.33, the direction of the PIM electric field with respect to the "radiation angle θ_{im}" is given.

BIBLIOGRAPHY

1. Bennett W. R., Rice S. O., Note on methods of computing modulation products. *Philosophical Magzine*, 1934, 9: 422–424.
2. Feuerstein E., Intermodulation products for V - law biased wave rectifier for multiple frequency input. *Quarterly of Applied Mathematics*, Apr 1957–Jan 1958, 15.
3. Brockbank R. A., Wass C. A. A., Non - linear distortion in transmission systems. *Electrical Engineers Part III Radio & Communication Engineering Journal of the Institution*, 1945, 92(55): 45–56.
4. Zhang S., Ge D., Wei B., The analysis and prediction of the power level of pim due to metallic contact nonlinearity at microwave frequencies. *Journal of Microwaves*, 2002, 18(4): 26–34.
5. Zhang S. Q., Wang, Calculation of pim amplitude and measurement of pim power level. *Journal of Changde Teachers University*, 2002, 14(3): 69–72.
6. Zhang S., Ge D., Analysis of general nature of intermodulation products based on fourier series method. *Chinese Journal of Radio Science*, 2005, 20(2): 265–268.
7. Zhang S. Q., Wu Q. D., Chen C., et al., Prediction of passive intermodulation power level at microwave frequencies. *Asia - Pacific Microwave Conference*. IEEE, 2008.
8. Wang C.-M., Wang G.-M., Zhang B., The analysis and calculation of the power level of pim due to mim based on power series. *Journal Of Air Force Engineering University(Natural Science Edition)*, 2008, 9(1):37–40.
9. Eng K.Y., Stern T. E., The order - and - type prediction problem arising from passive intermodulation interference in communications satellites. *IEEE Transactions on Communications Systems*, 1981, 29(5): 549–555.
10. Eng K. Y., Yue O. C., High - order intermodulation effects in digital satellite channels. *IEEE Transactions on Aerospace and Electronic Systems*, 1981, 17(3): 438–445.
11. Boyhan J. W., Ratio of Gaussian PIM to two - carrier PIM. *IEEE Transactions on Aerospace and Electronic Systems*, 2000, 36(4): 1336–1342.
12. Abuelma'Atti M. T., Prediction of passive intermodulation arising from corrosion. *IEEE Proceedings: Science, Measurement and Technology*, 2003, 150(1): 30–34.
13. Abuelma'Atti M. T., Large signal analysis of the Y - fed directional coupler. *Frequenz*, 2008, 62(11): 288–292.
14. Vicente C., Hartnagel H. L., Passive - intermodulation analysis between rough rectangular waveguide flanges. *IEEE Transactions on Microwave Theory and Techniques*, 2005, 53(8): 2515–2525.

15. Johannes, R., Aravind, R., An-Dreas, C., et al., phenomenological modeling of passive inter-modulation (Pim) due to electron tunneling at metallic contacts. *International Microwave Symposium Digest.* IEEE, 2006.

16. Shitvov A. P., Zelenchuk D. E., Schuchinsky A. G., et al., Studies On Passive Intermodulation Phenomena In Printed And Layered Transmission Lines. *High Frequency Postgraduate Student Colloquium, 2005,* IEEE, 2005.

17. Zelenchuk D., Shitvov A. P., Schuchinsky A. G., et al., Passive intermodulation on microstrip lines. *European Microwave Conference.* IEEE, 2007.

18. Zelenchuk D. E., Shitvov A. P., Schuchinsky A. G., et al., Passive inter-modulation in finite lengths of printed microstrip lines. *IEEE Transactions on Microwave Theory and Techniques,* 2008, 56 (11): 2426–2434.

19. Henrie J., Christianson A., Chappell W. J., Prediction of passive intermodulation from coaxial connectors in microwave networks. *IEEE Transactions on Microwave Theory and Techniques,* 2008, 56 (1): 209–216.

20. Bolli P., Naldini A., Pelosi G., et al., A time domain physical optics approach to passive inter-modulation scattering problems. *Antennas & Propagation Society International Symposium.* IEEE, 1999.

21. Selleri S., Bolli P., Pelosi G., A time - domain physical optics heuristic approach to passive intermodulation scattering. *IEEE Transactions on Electromagnetic Compatibility,* 2001, 43 (2): 203–209.

22. Bolli P., Selleri S., Pelosi G., Passive intermodulation on large reflector antennas. *IEEE Antenna's And Propagation Magazine,* 2002, 44 (5): 13–20.

23. Selleri S., Bolli P., Pelosi G., Automatic evluation of the non - linear model coefficients in passive intermodulation scattering via genetic algorithms. *IEEE Antennas and Propagation Society International Symposium,* 2003: 390–393.

24. Selleri S., Bolli P., Pelosi G., Genetic algorithms for the determination of the nonlinearity model coefficients in passive intermodulation scattering. *IEEE Transactions on Electromagnetic Compatibility,* 2004, 46 (2): 309–311.

25. Selleri S., Bolli P., Pelosi G., Some insight on the behaviour of heuristic PIM scattering models for td - po analysis. *IEEE Antennas and Propagation Society International Symposium,* 2004.

26. Lojacono R., Mencattini A., Salmeri M., et al., simulation of the effects of the residual low level pim to improve payload design of communication satellites. *IEEE Aerospace Conference.* IEEE, 2005.

27. Ishibashi D., Kuga N., Analysis Of 3rd - order passive intermodulation generated from metallic materials. *Asia - Pacific Microwave Conference.* IEEE, 2008.

28. You J. W., Wang H. G., Zhang J. F., et al., Accurate numerical analysis of nonlinearities caused by multipactor in microwave devices. *IEEE Microwave and Wireless Components Letters,* 2014, 24 (11): 730–732.

29. You J. W., Zhang J. F., Gu W. H., et al., Numerical analysis of passive inter-modulation arisen from nonlinear contacts in HPMW devices. *IEEE Transactions On Electromagnetic Compatibility,* 2017, 60 (5): 1–11.

30. You J. W., Cui T. J., Modeling nonlinear phenomena of multipactor and passive intermodulation. *Microwave Conference.* IEEE, 2017.

31. Wang Y. L., Wang Y. W., Nai C. X., et al., 2D modelling and simulation of dc resistivity using comsol. *2012 Second International Conference On Instrumentation，Measurement，Computer, Communication And Control* (IMCCC). IEEE, 2012.

32. Li S.-J., Wang H.-Q., Zhao W.-Y., Meng W.-J., Wen H., Multiphysics coupling simulation modeling methods based on COMSOL. *Mechanical Engineering & Automation,* 2014, 185 (4): 19–20, 23.

33. Wang X. B., Cui W. Z., Wang J. Y., et al., 2D PIM Simulation Based On COMSOL. Piers 2011 Marrakesh Proceedings. The Electromagnetics Academy, 2011: 181–185.

34. Gu W., *Research on Numerical Method of Passive Intermodulation*. Nanjing: Southeast University, 2016.
35. Wu D. W., Xie Y. J., Kuang Y., et al., Prediction of passive intermodulation on grid reflector antenna using collaborative simulation: Multiscale equivalent method and nonlinear model. *IEEE Antenna and Propagation*, 2018, 66 (3): 1516–1521.
36 Jiang J., *On Mechanical-Thermal-Electrical Effect of Passive Intermodulation in Microwave Devices*. Xi'An: Xidian University, 2016.
37. Liu Y., Mao Y. R., Xie Y. J., et al., Evaluation of passive intermodulation using full - wave frequency - domain method with nonlinear circuit model. *IEEE Transactions on Vehicular Technology*, 2016, 65 (7): 5754–5757.
38. Mao Y. R., Liu Y., Xie Y. J., et al., Simulation of electromagnetic performance on grid reflector antennas: Three - dimensional grid structures with lumped boundary conditions. *IEEE Transactions on Antennas & Propagation*, 2015, 63(10): 4599–4603.
39. Wu D., Xie Y., Kuang Y., et al., Prediction of passive intermodulation on grid reflector antenna using collaborative simulation: Multiscale equivalent method and nonlinear model. *IEEE Transactions on Antennas and Propagation*, 2018, 66(3): 1516–1521.
40. Mao Y. R., Xie Y. J., Analysis of electromagnetic multi-scale structure and non-linear effects. *Journal of Beijing University of Aeronautics and Astronautics*, 2015, 41(10): 1848–1852 (In Chinese).
41. Mao Y.-R., Liu Y., Xie Y.-J., Cheng Z.-H., Wang X.-B, Li Y., Numerical analysis of passive intermodulation due to metallic contact nonlinearity. *Chinese Journal of Electronics*, 2015, 43(6): 1174–1178.

Passive Intermodulation Localization Technology for Microwave Components

Xiang Chen, Wanzhao Cui, and Lixin Ran

CONTENTS

DOI: 10.1201/9781003269953-4

4.1 GENERAL OVERVIEW

The generation mechanism of passive intermodulation (PIM) is very complex. There are many nonlinear sources in the actual microwave components and systems, and under the high-power carrier excitation, each nonlinear source may generate PIMs, which are eventually superimposed to form PIM interference output. Because the physical causes of PIM are complex and diverse and exhibit unpredictability under the action of multiple factors, the existence of PIM hazards cannot be completely avoided, and there are often multiple possible PIM hazards or occurrence points within a microwave component (or subsystem). These problems are magnified by the special conditions in the satellite-based environment, thus appearing more serious, usually only in a variety of combinations of test conditions, and prolonged observation is possible to obtain reliable data, and very often even if the PIM component is measured, it is difficult to effectively locate its occurrence point, which can seriously affect the product development cycle. If you can accurately find the main key PIM sources, you can target suppression, and then significantly improve the efficiency of solving the PIM interference, which is undoubtedly one of the very effective ways to solve the PIM problem. In summary, it is of great importance to carry out research on PIM detection and localization technology for microwave components.

At present, there are no effective methods reported publicly about the localization of arbitrary multipoint PIM sources. The main difficulty is that multiple PIM sources are coupled and superimposed on each other, and it is difficult to obtain their position information directly through the test results. The key to achieve PIM localization lies in how to extract the physical location information through theoretical modeling and innovative algorithms based on the phase and amplitude information of PIM occurrence points on the basis of high-sensitivity PIM detection. Meanwhile, PIM localization is carried out based on the test system, while most of the actual systems and components are narrow-band devices, which cannot provide sufficient sampling bandwidth, thus further increasing the complexity of theoretical calculation. Therefore, another key to realize PIM localization is how to get rid of the actual hardware bandwidth limitation through an effective error optimization method to realize the transition from theory to application.

4.2 APPLICATION BACKGROUND AND RESEARCH STATUS OF PASSIVE INTERMODULATION LOCALIZATION TECHNOLOGY

4.2.1 PIM Source Localization Technology for Closed Structure

PIM localization can be classified into localization in the closed cavity structures and open structures according to the form of the DUT (device under test). There are few literature studies on PIM location of closed cavity structure. This is mainly because the PIM localization method of closed cavity structure basically depends on the inversion of phase (or standing wave) properties to obtain the physical length. Although this method may be applicable to simple structures (such as transmission lines), in a complex structure (or multiple complex structures) combined into a system, its internal electromagnetic wave transmission path and phase characteristics will be very complicated. The information generated by phase (or standing wave) inversion cannot directly correspond to the physical length, but depends on different electromagnetic structure models, so it is difficult to propose a universal PIM localization method for the cavity structure.

The PIM measurement system alone cannot provide information on the location of the PIM occurrence point, and there are three main methods for locating the PIM occurrence point and hidden trouble point for closed cavity structures as follows:

1. Knock test. A small rubber hammer is used to knock on each component and cable connector of the system under test, and the PIM level is continuously monitored. Under this interference, the PIM level will fluctuate greatly if there are hidden defects in the PIM.

2. Progressive measurement. The system under test is assembled (or disassembled) step by step, and the PIM measurement is carried out for the partially assembled system at each stage, so as to reduce the ambiguity of position information and determine the components (components) in the system where PIM occurs.

3. Manual inspection of possible hidden trouble points.

The above methods all require manual participation and can only make rough judgments, which cannot accurately achieve the localization of multiple PIM occurrence points.

Since 2017, we together with Lixin Ran of Zhejiang University systematically and comprehensively studied the method flow of PIM source localization for closed structure microwave components, put forward a new theoretical method of PIM localization based on wave vector k-space Fourier transform and combined with the actual situation, gave the targeted error optimization algorithm and achieved a certain degree of experimental verification on results. This is the most in-depth and valuable research achievement in PIM localization of closed structures at home and abroad, which will be introduced in detail later in this book.

4.2.2 PIM Source Localization for Open Structure

Most of the published literature on PIM localization methods focuses on PIM localization in the open structures, and there are two main methods: near-field scanning and additional excitation.

The principle of PIM localization by near-field scanning is that first, carrier signals are sent out by two transmitting antennas to excite the open structure to be tested; then, the PIM signals generated by the DUT are received by a highly directional receiving antenna; finally, a mechanical scanning mechanism is used to make the receiving antenna complete the scanning of a certain plane. Some scholars have used this method to achieve the scanning test of the antenna darkroom wall at 1.5 GHz. Microwave holographic imaging method can also be used in near-field scanning, the reflective surface is excited through the two transmitting antennas, the sampling of the scattered field is performed at the PIM frequency by plane scanning the near-field in the reflective surface and then the distribution

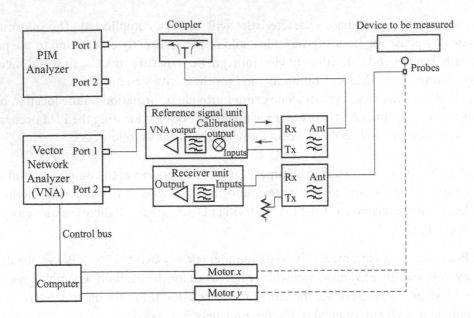

FIGURE 4.1   Block diagram of the near-field scanning system.

of the PIM field on the reflective surface is derived by using the microwave holographic imaging method.

A more representative work on the near-field scanning method is the one done by Antti V. Raisanen and Sami Hienonen et al. at Helsinki University of Technology, Finland. They designed the near-field scanning method, whose system block diagram is shown in Figure 4.1. The idea is that the field intensity at the PIM frequency of a DUT with open structure is scanned using a probe in its near field, and the distance of the probe from the DUT is required to be less than ten times the wavelength in order to obtain a high sensitivity and resolution. They used this method to detect the amplitude and phase of the third-order PIM in GSM 900 MHz band, with which they realized the PIM localization within the scanning range of $0.3 \times 1 \, m^2$. In their published paper, they mentioned that they can use this scanning method to achieve the PIM source localization of antenna and micro-strip line under two 20 W input conditions in the sensitivity range of $-110 \sim -80$ dBm.

The main limitation of the near-field scanning method is that the probe itself generates PIM, which has an impact on the sensitivity of the detection. The PIM generated by the probe depends mainly on the structure of the probe as well as the DUT.

The method of using applied excitation to achieve open structure PIM localization was proposed by John. C. Mantovani et al. in 1987. The core idea is that the PIM is generated by excitation of the DUT (e.g., MIM junction) by two carriers, while the DUT is excited by a focused high-power acoustic wave beam, and the high-power acoustic signal causes the DUT to vibrate, thus affecting its PIM characteristics. The specific implementation process is that the acoustic wave frequency is modulated to the sideband of the PIM product, and then PIM localization is performed through the detection of the sideband component. The principle of acoustic excitation PIM localization is shown schematically in Figure 4.2.

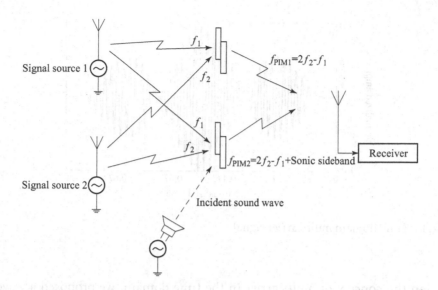

FIGURE 4.2 Schematic diagram of the principle of acoustic excitation PIM localization.

4.3 THEORETICAL METHOD AND IMPLEMENTATION FRAMEWORK OF PIM LOCALIZATION BASED ON WAVE VECTOR K-SPACE FOURIER TRANSFORM

4.3.1 The Theory of PIM Localization Algorithm Based on Wave Vector k-space Fourier Transform

Multicarrier modulation technology is widely used in modern wireless communication systems. In a typical communication application, the time-domain multicarrier signal is synthesized by N carriers with amplitude V_i, phase φ_i and the frequency difference of its carriers $\Delta f k$, and the multicarrier signal can be written as

$$v(t) = \sum_{i=1}^{N} V_i \exp\left(j\left(2\pi\left(f_0 + \sum_{k=1}^{i-1} \Delta f_k \right)t + \varphi_i \right) \right) \tag{4.1}$$

where f_0 represents the initial frequency.

If the initial phase is zero ($\varphi_i = 0$), the frequency interval is the same ($\Delta f_k = \Delta f$), and the carrier is distributed with equal amplitude ($V_i = V_0$); then the formula (4.1) can be simplified as

$$v(t) = V_0 \sum_{i=1}^{N} \exp\left(j\left(2\pi\left(f_0 + (i-1)\Delta f \right)t + \varphi_i \right) \right)$$

$$= V_0 \frac{\sin(\pi N \Delta f t)}{\sin(\pi \Delta f t)} \exp\left(j 2\pi f_0 t \right) \tag{4.2}$$

Figure 4.3 shows the time domain multicarrier signal composed of six channels with equal amplitude and equal frequency interval. T is the period of the multicarrier signal.

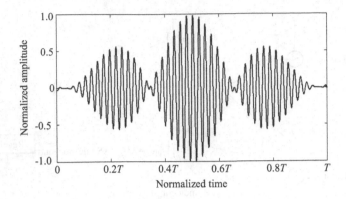

FIGURE 4.3 Time domain multicarrier signal.

Based on the concept of multicarrier in the time domain, we proposed a wave vector k-space Fourier transform localization algorithm to achieve PIM source localization, and the specific steps are as follows:

Step 1: Two controllable coherent excitation signal sources and one PIM reference signal source share the same reference signal source. The two controllable coherent excitation signal sources, respectively, generate single-frequency controllable coherent radio frequency signals as the transmission carrier with frequencies f_1 and f_2, respectively, and the signals emitted by the two controllable coherent radio frequency signal sources are amplified, combined and injected into the DUT; when the total power of the injected two signals is greater than the trigger threshold of the DUT, the DUT generates the actual PIM signal, while the PIM reference signal source generates a "virtual" PIM reference signal with the frequency $f_3 = \alpha f_1 + \beta f_2$, where α and β are the order parameters of the PIM signal, which correspond to the order of the PIM signal that actually needs to be measured.

Step 2: At the receiving end, the actual PIM signal and the "virtual" PIM reference signal are filtered by the corresponding filters, and the phase difference between the actual PIM signal and the "virtual" PIM reference signal is obtained as Φ through the phase comparator, and the amplitude value A of the actual PIM signal is obtained through the time domain Fourier transform measurement.

Step 3: Linearly increase the power of the two controllable coherent RF signals, and judge whether a new PIM signal is generated by the change of the phase difference Φ of the PIM signal under different power, so as to obtain a new PIM signal trigger threshold.

Step 4: Set the power of the two controllable coherent excitation signal sources as the new PIM signal trigger threshold.

Step 5: Repeat the above steps under linearly changing the frequency value of any one of the two controlled coherent RF signals at equal intervals to obtain the amplitude value A_n and phase difference value Φ_n of the multigroup PIM signal.

Step 6: The measured amplitude value A_n and phase difference value Φ_n of the PIM signal are synthesized into a vector signal in the form of $A_n e^{j\phi_n}$.

Assuming that there are N PIM sources in the uniform waveguide cavity to be tested, and the distances from the incident port are $x_1, x_2, ..., x_N$, for any PIM source, the vector signal of the i-th PIM source can be obtained according to the above localization steps in the form of

$$s_i(k_{PIM}) = A_i \exp(j\Delta\Phi_i(k_{PIM}))\qquad(4.3)$$

where A_i represents the PIM signal amplitude at a specific order; $\Delta\varphi_i$ (kPIM) represents the phase shift of the PIM signal at a specific order generated from the incident port x_i to the device port. It should be noted that as the frequency of the coherent excitation signal source changes each time, the PIM signal wave vector k_{PIM} also changes linearly.

Finally, according to the vector synthesis principle, the vector signal synthesized by the measured PIM amplitude value A_n and the phase difference value Φ_n is equal to the superposition of all the PIM source vector signals, that is,

$$A(k_{PIM})\exp(j\Phi(k_{PIM})) = \sum_{i=1}^{N} A_i \exp(j\Delta\Phi_i(k_{PIM}))\qquad(4.4)$$

Write formula (4.4) as a function:

$$F(k_{PIM}) = \sum_{i=1}^{N} A_i \exp\left(j\left(m(\varphi_1+\Delta\varphi_1)+n(\varphi_2+\Delta\varphi_2)+2k_{PIM}x_i+\Delta\varphi_{MIM}\right)\right)\qquad(4.5)$$

where Φ_1 and Φ_2 represent the initial phase of the two test signals; $\Delta\varphi_{MIM}$ represents the fixed phase shift of the PIM signal at the point of occurrence.

It can be seen that the function $F(k_{PIM})$ is a multicarrier signal in the wave vector k-space. Therefore, it can be considered that the PIM products generated by multiple PIM sources inside the transmission line will form a "multicarrier" in the wave vector k space after being superimposed at the device port. However, for the specified order of PIM products, the variable of the multicarrier is the wave vector, not the time, and its corresponding "frequency" is the distance of each PIM source from the port.

Based on the above understanding, via Fourier transform, the function $F(k_{PIM})$ can be transformed in the wave vector k-space. According to Equation 4.5, the complex signal after the Fourier transform of wave vector k space, the "spectrum" obtained is a series of impulse strings, whose horizontal coordinate values are $x_1, x_2, ..., x_N$, that is, the location information of the PIM source. Through the above method, it is possible to theoretically realize the localization of any multiple PIM sources, and the simplified flow of the algorithm is shown in Figure 4.4.

4.3.2 PIM Localization Implementation Architecture Based on Wave Vector k-space Fourier Transform

According to the localization theory in Section 4.3.1, in the PIM localization based on wave vector k-space Fourier transform, the amplitude and phase of the "virtual" PIM reference signal generated artificially are compared with those of actual PIM signal generated by the DUT to obtain the amplitude and phase information of the actual PIM signal, so the localization algorithm should be based on the PIM measurement. Therefore, the

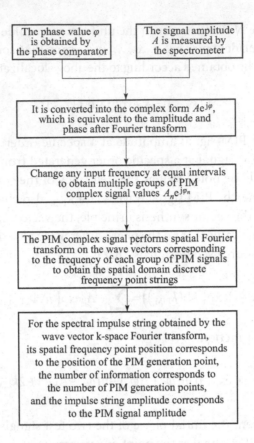

The phase value φ is obtained by the phase comparator

The signal amplitude A is measured by the spectrometer

It is converted into the complex form $Ae^{j\varphi}$, which is equivalent to the amplitude and phase after Fourier transform

Change any input frequency at equal intervals to obtain multiple groups of PIM complex signal values $A_n e^{j\varphi_n}$

The PIM complex signal performs spatial Fourier transform on the wave vectors corresponding to the frequency of each group of PIM signals to obtain the spatial domain discrete frequency point strings

For the spectral impulse string obtained by the wave vector k-space Fourier transform, its spatial frequency point position corresponds to the position of the PIM generation point, the number of information corresponds to the number of PIM generation points, and the impulse string amplitude corresponds to the PIM signal amplitude

FIGURE 4.4   Flow chart of PIM source localization method based on wave vector k-space Fourier transform.

implementation of the localization algorithm should be based on the PIM measurement, and the block diagram of the implementation of the PIM localization theory is shown in Figure 4.5. Three vector signal sources sharing the same reference clock are used. By precisely controlling the initial phases of excitation signal sources 1 and 2, according to the order of PIM to be measured, signal source 3 generates the corresponding PIM frequency and initial phase, so as to generate a "virtual" artificial PIM reference source, which is located at the output port of signal source 3. Obviously, since the same reference clock is used, signal source 1, signal source 2, and signal source 3 are all coherent. The signals with frequencies f_1 and f_2 and initial phases Φ_1 and Φ_2 output, respectively, by signal source 1 and signal source 2 are amplified by power amplifiers and then are combined into one signal by a combiner, which reaches the DUT through a duplexer, thus generating a PIM signal inside the DUT, from which the reflected PIM signal is separated by the duplexer, and the required actual PIM signal is obtained after the reflected PIM signal passes through a filter. The signal source 3 generates a "virtual" artificial PIM reference source with frequency $mf_1 \pm nf_2$, where m and n are positive integers and initial phase Φ_3. Finally, the actual PIM signal and the virtual PIM signal are sent to the phase comparator for phase comparison to obtain the phase difference between the two signals.

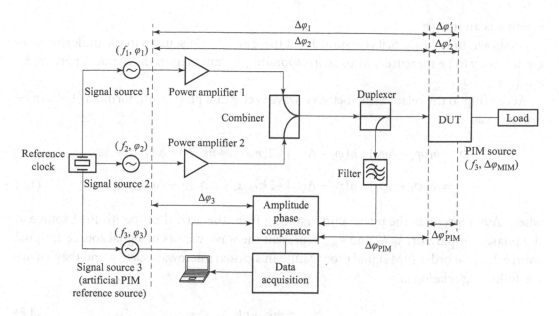

FIGURE 4.5 Block diagram of the implementation architecture of PIM localization theory.

The phase shifts of signal source 1 and 2 from the source to the DUT are denoted as $\Delta\varphi_1$ and $\Delta\varphi_2$, respectively, and the reflected PIM product generated by the DUT is passed through the filter to obtain the actual PIM signal of a specific order. Since the phase delay of the duplexer and filter can be set and adjusted, the phase difference of the actual PIM signal from the duplexer to the phase comparator is known and is denoted as $\Delta\varphi_{PIM}$. The phase of the virtual signal generated by signal source 3 is φ_3; the phase shift of the virtual PIM signal from the source to the phase comparator is known and is denoted as $\Delta\varphi_3$. What needs to be solved is the unknown phase $\Delta\varphi'_{PIM}$ of which the PIM signal generated by the DUT is reflected to the port connected with the diplexer. The phase difference between the two signals obtained after signal processing is $\Delta\varphi_m$. When there are multiple PIM sources, the phase difference $\Delta\varphi_m$ with the "virtual" PIM reference source can be measured by the phase comparator. $\Delta\varphi_m$ is the phase difference between the superposition phase of multiple PIM signals of the same order from all positions and the "virtual" PIM signal.

$m+n$ is the order of the PIM signal, and the phase of the corresponding frequency PIM signal can be measured by adjusting the values of m and n. The initial phase of the PIM signal at the PIM generation point is corresponding to the order, and the initial phase corresponding to the m+n order PIM signal is $m(\varphi_1+\Delta\varphi_1+k_1x)+n(\varphi_2+\Delta\varphi_2+k_2\chi)+\Delta\varphi_{MIM}$, k_1 and k_2 are the two wave vectors of the test signal, and $\Delta\varphi_{MIM}$ is the fixed phase shift of the PIM signal at the point of occurrence. Therefore, $\Delta\varphi_m$ can be obtained as

$$\Delta\varphi_m = m\left(\varphi_1 + \Delta\varphi_1\right) + n\left(\varphi_2 + \Delta\varphi_2\right) + 2\Delta\varphi'_{PIM} + \Delta\varphi_{PIM} + \Delta\varphi_{MIM}$$

$$-\varphi_3 - \Delta\varphi_3 + 2d\pi \tag{4.6}$$

where d is an integer.

As shown in Figure 4.6, it is assumed that there are N PIM sources in the uniform waveguide cavity to be measured, whose corresponding distances from the incident port are x_1, x_2, ..., x_N.

According to the relationship between wave vector and phase shift, formula (4.6) can be rewritten as

$$\Delta\Phi_i = m(\varphi_1 + \Delta\varphi_1) + n(\varphi_2 + \Delta\varphi_2) + 2(mk_1x_i + nk_2x_i) + \Delta\varphi_{\text{MIM}} + \Delta\varphi_{\text{PIM}}$$

$$= m(\varphi_1 + \Delta\varphi_1) + n(\varphi_2 + \Delta\varphi_2) + 2k_{\text{PIM}}x_i + \Delta\varphi_{\text{MIM}} + \Delta\varphi_{\text{PIM}} \tag{4.7}$$

where $\Delta\Phi_i$ represents the phase shift resulting from the arrival of the ith PIM source at the phase comparator; k_1, k_2 and k_{PIM} represent the wave vectors of signal source 1, signal source 2, $m+n$ order PIM signal propagating in a closed microwave cavity, and they satisfy the following relationship:

$$k_{\text{PIM}} = mk_1 + nk_2 \tag{4.8}$$

In practical applications, the microwave cavity to be tested is often nonuniform, and there are multiple PIM locations in it. Figure 4.7 shows a nonuniform microwave cavity PIM transmission model, where there are N PIM sources, and their corresponding positions from the entrance port are x_1, x_2, ..., x_N. There are P subsegments in the microwave cavity under test. The uniform subsegment, for a PIM signal of a certain order, its wave vectors are k_{PIM1}, k_{PIM2}, ..., k_{PIMP}. Assuming that the i-th PIM source falls on the k-th uniform subsegment, the phase of the $m+n$-th order PIM signal reaching the phase comparator is

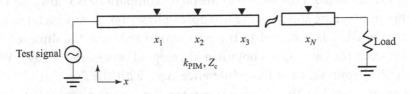

FIGURE 4.6 Uniformly closed cavity model. (▼—PIM source; Zc—Specific impedance of transmission line.)

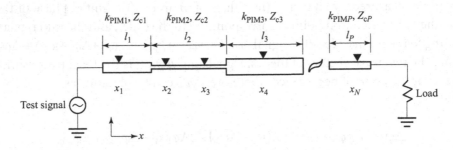

FIGURE 4.7 Nonuniform microwave cavity PIM transmission model. (▼—PIM source.)

$$\Delta\Phi_i = m\left(\varphi_1 + \Delta\varphi_1\right) + n\left(\varphi_2 + \Delta\varphi_2\right) + 2\left(\sum_{j=1}^{k-1} k_{\text{PIM}j}l_j + k_{\text{PIM}j}\left(x_i - \sum_{j=1}^{k-1} l_j\right)\right)$$
$$+ \Delta\varphi_{\text{MIM}} + \Delta\varphi_{\text{PIM}}$$

$$(4.9)$$

where $\Delta\Phi_i$ is the phase shift produced by the i-th PIM source reaching the phase comparator, $k_{\text{PIM}j}$ is the wave vector of the uniform transmission subsection of the $m+n$-order PIM signal in the closed microwave cavity and I_j is the length of the j-th uniform propagation subsegment in the closed microwave cavity.

Similarly, on the jth uniformly propagating subsegment of the closed microwave cavity, the wave vectors of signal source 1, signal source 2 and $m+n$th order PIM signals propagating in the closed microwave cavity satisfy the following relationship

$$k_{\text{PIM}j} = mk_{1j} + nk_{2j} \qquad (4.10)$$

where k_{1j}, k_{2j} and k_{PIMj} represent the signal source 1, the signal source 2 and the $m+n$ order PIM signal that propagates the wave vector of the subsegment uniformly in the jth segment in the closed microwave cavity.

Therefore, for a nonuniform microwave cavity, an "average" wave vector can be obtained as

$$\bar{k}_{\text{PIM}} = \left(\sum_{j=1}^{P} k_{\text{PIM}j}l_j\right) / L \qquad (4.11)$$

where P denotes the number of all subsections of the nonuniform microwave cavity; L denotes the physical length of the nonuniform microwave cavity.

Therefore, the nonuniform microwave cavity can be equivalent to a uniform microwave cavity with an average wave vector.

When the microwave cavity to be tested is a nonuniform and complex microwave cavity, the PIM source position obtained according to the method in Section 4.3.1 needs to be processed later. From Equations 4.10 and 4.11, the actual position of the PIM source can be obtained as

$$x_i = \sum_{j=1}^{S_i-1} l_j + \left(\bar{k}_{\text{PIM}}x_i' - \sum_{j=1}^{S_i-1} k_{\text{PIM}j}l_j\right) / k_{\text{PIMS}} \qquad (4.12)$$

where S_i represents the transmission subsegment where the i-th PIM source is located; x'_i represents the position of the theoretical PIM source obtained by the Fourier transform of the wave vector k-space.

4.3.3 Application Scope and Conditions of Wave Vector k-Space Fourier Transform-Based PIM Localization Theory Algorithm

Based on the wave vector k-space Fourier transform localization algorithm described in the previous section, the following requirements must be met in specific implementation:

1. In addition to the nonlinearities that cause PIM, the DUT is also a linear device with well-matched input/output ports within the measurement bandwidth.

2. The measurement system can ensure that the phase of the PIM signal synthesized by the specified order of PIM components is measured at the device port.

3. The measurement system bandwidth can guarantee the k-domain discrete sampling interval and sampling length for the synthesized PIM signal.

4. The measurement system needs to filter out the PIM signal at other orders than the specified order PIM signal and excitation carrier signal.

5. The amplitude characteristics of each PIM generation point remain stable within the measurement bandwidth.

In addition to the above conditions, according to the principle of the algorithm, the following parameter conditions need to be satisfied:

a. Wave vector sampling rate

In the wave vector k-space Fourier transform-based PIM localization algorithm, the applicability and usage conditions of the Fourier transform itself must first be satisfied because of the need for the Fourier transform.

The sampling accuracy f_s of the Fourier transform must meet the Nyquist sampling theorem to prevent spectral aliasing. From equation (4.5), we can obtain

$$f_s = \frac{1}{\Delta k_{\text{PIM}}} \geq 2f_{\text{max}} \qquad (4.13)$$

where Δk_{PIM} represents the sampling interval in the wave vector k-space, and f_{max} represents the highest frequency in the wave vector k-space spectrum, i.e., the farthest PIM source distance from the incident port.

Assuming that the length of the microwave cavity device to be tested is L, then

$$f_s \geq 4L \qquad (4.14)$$

b. Wave vector k-space spectral resolution

In the Fourier transform, the distance between two effective spectral lines is defined as the spectral resolution. As mentioned before, after the Fourier transform of the wave vector k-space of the PIM complex signal, the obtained spectrum is a series of impulse strings whose horizontal coordinate values are $x_1, x_2, \ldots, x_N$, i.e., the location information of the PIM source. Since the sampling length of wave vector sampling is limited, in order to effectively distinguish the spectrum impulse strings in wave vector k-space, the spectrum resolution must be smaller than the minimum interval of spectrum impulse strings. Assuming that the minimum interval of the spectral impulse string is $\Delta x_{\min}$, the corresponding spectral resolution in the wave vector k-space should satisfy

$$\Delta x_{\min} > f_s / M = 1 / (\Delta K + \Delta k_{PIM})$$ (4.15)

where M represents the number of sampling points in the k-space of the wave vector, and ΔK represents the sampling length of the wave vector k-space.

c. Wave vector k-space multicarrier period condition

The vector signal constructed by Equation 4.5 is obtained by superimposing multiple PIM sources in the transmission line at the device ports, which will form a "multicarrier" in the wave vector k-space. Therefore, for the Fourier transform spectrum of periodic signals, the sampling length needs to be at least one period to reflect the periodic signal information on the spectrum effectively. As an example, we can rewrite Equation 4.5 for a multicarrier of equal frequency interval in wave vector k-space as

$$F(k_{PIM})$$

$$= \sum_{i=1}^{N} A_i \exp\left(j\left(m(\varphi_1 + \Delta\varphi_1) + n(\varphi_2 + \Delta\varphi_2) + 2k_{PIM}(x_0 + (i-1)\Delta x) + \Delta\varphi_{MIM} \right) \right)$$ (4.16)

where x_0 represents the distance of the position of the first PIM point from the incident port; Δx represents the interval of two adjacent PIM points.

As a result, the multicarrier period of the wave vector k-space multicarrier described by Equation 4.16 is $\pi/\Delta x$. If the resolution of the position interval of the two PIM sources is required to be $\Delta x = \alpha \lambda g$, ($\alpha$ is the coefficient of wavelength, and λg is the wavelength at a particular operating frequency of a component to be measured), the sampling length of its wave vector k-space should at least satisfy

$$\Delta K \geq \pi / (a\lambda_g)$$ (4.17)

Assuming that λg is 1m, in order to obtain a resolution of 0.5 λg or 0.1λg, at least 300MHz and 1.5 GHz are required for the scanning frequency of the control excitation source. Therefore, the wave vector k-space Fourier transform PIM localization algorithm for the resolution of similar PIM sources (or the localization of PIM source positions under small-scale microwave components) is often limited by the bandwidth, so the algorithm implementation is to some extent limited by the bandwidth of the PIM measurement system.

As a result, through the innovative use of an equivalent k-space multicarrier concept, the wave vector k-space Fourier transform PIM localization algorithm can achieve the simultaneous localization of multiple PIM sources within the closed microwave cavity and can reflect the propagation and reflection of electromagnetic waves inside the microwave component to be measured to a certain extent. In addition, by scanning the PIM excitation signal source in a certain bandwidth to achieve sampling, and combined with vector signal construction, the spectrum information obtained after the Fourier transform has the ability to remove phase ambiguity.

4.4 ERROR OPTIMIZATION METHOD FOR PIM LOCALIZATION AND SIMULATION EXPERIMENTAL VERIFICATION

4.4.1 Wave Vector K-space Inverse Problem Optimization PIM Localization Algorithm

In the case of multiple PIM source localization, the wave vector k-space Fourier transform localization algorithm described above is constrained by the Fourier transform accuracy, which often requires a wide bandwidth, and thus, the algorithm is usually limited in narrow-band systems. In addition, according to the mechanism of PIM, in the actual microwave components to be measured, the locations where PIM may be generated are mainly bolts, connections and other discontinuous structural locations, as well as rusted parts. Therefore, for the microwave components to be measured, all the possible locations of PIM effect can be predicted to make it a known quantity. At this point, the localization of PIM is transformed into the determination of the actual location of PIM occurrence among all possible PIM sources. Due to the introduction of a priori knowledge, the nonlinearity of Equation (4.16) is reduced, so that it becomes feasible to achieve PIM localization in narrow-band bandwidth. Further, this chapter proposes a wave vector k-space inverse problem optimization localization algorithm to achieve PIM source localization under narrowband measurement conditions, which includes the following steps.

Step 1: Under the same reference signal source, two coherent signal sources are injected into the microwave cavity to be measured, and PIM signals are generated at known multiple possible PIM sources (which can be called PIM hidden points). Multiple PIM signals at a specific order are reflected and superimposed at the incident port, and their superimposed synthesis can be regarded as a multicarrier signal, and the amplitude and phase of the superimposed signal can be obtained by an amplitude phase comparator.

Step 2: Change the frequency of one of the coherent RF signals to obtain the amplitude and phase of multiple superimposed signals, and then build a complex system of equations:

$$
\begin{cases}
A_1 e^{j\Delta\Phi_{11}} + A_2 e^{j\Delta\Phi_{12}} + \cdots + A_n e^{j\Delta\Phi_{1n}} = A(1)e^{j\Delta\Phi(1)} \\
A_1 e^{j\Delta\Phi_{21}} + A_2 e^{j\Delta\Phi_{n}} + \cdots + A_n e^{j\Delta\Phi_{20}} = A(2)e^{j\Delta\Phi(2)} \\
A_1 e^{j\Delta\Phi_{m1}} + A_2 e^{j\Delta\Phi_{m2}} + \cdots + A_n e^{j\Delta\Phi_{m}} = A(m)e^{j\Delta\Phi(m)}
\end{cases}
\tag{4.18}
$$

where A_n represents the amplitude of the PIM signal generated at the n-th PIM hidden point; $\Delta\Phi_{mn}$ represents the phase of the PIM signal generated at the n-th PIM hidden point at the m-th measurement, which can be represented as

$$\Delta\Phi_{mn} = 2x_n k_{PIMm} + \Delta\varphi_{MIM} \tag{4.19}$$

where x_n represents the position of the n-th PIM hidden point from the port; $kPIMm$ represents the wave vector constant at the m-th measurement; and $\Delta\varphi_{MIMn}$ represents the initial phase of the PIM signal generated by the nth PIM hidden point; $A(m)$ represents the amplitude of the synthesized signal measured by the receiver at the m-th measurement; and $\Delta\Phi(m)$ represents the phase of the synthesized signal measured by the receiver at the m-th measurement.

Step 3: Construct the optimization objective function:

$$F_{opt} = \min \sum_{i=1}^{M} \left\| \sum_{j=1}^{N} A_j e^{j\Delta\Phi_y} - A(i)e^{j\Delta\Phi(i)} \right\| \tag{4.20}$$

where $A_1, A_2, ... A_n$ and $\Delta\varphi_{MIM1}, \Delta\varphi_{MIM2}, ..., \Delta\varphi_{MIMn}$ are the optimization parameter variables. The initial values of the optimized parameter variables are set, and the most rapid descent method is used to find the optimal solution parameters.

After locating the PIM hidden point, the magnitude threshold condition should be set to determine whether the corresponding PIM hidden point really generates PIM. When the magnitude optimization parameter is lower than the threshold, the corresponding location is considered to have not generated PIM; when the magnitude optimization parameter is higher than the threshold, the corresponding location can be considered to have really generated PIM.

In the first step, multiple PIM signals of a specific order include PIM signals at 3rd, 5th, 7th or higher order.

In the second step, the frequency value of any one of the two controllable coherent radio frequency signals is changed at equal intervals, and the number of measurement samples should be greater than or equal to twice the number of all hidden PIM points. Moreover, the locations of all PIM hidden danger points are known.

In the third step, the sample data amplitude is normalized data to ensure that the data optimization search area is unified.

The flow block diagram of the optimal localization algorithm for the wave vector k-space inverse problem is shown in Figure 4.8. It is worth noting that this algorithm reduces the nonlinearity of the optimization objective function by obtaining a priori knowledge (i.e., predetermining all possible PIM hidden point location information), and then uses a local optimization algorithm to obtain its extreme value point. As for the complex microwave component structure, it may not be possible to determine all the PIM hidden point locations, or there are too many optimization parameter variables, when local optimization may not be achieved, and other more effective error optimization methods can only be used while seeking to expand the measurement bandwidth.

4.4.2 PIM Localization Algorithm Based on Matrix Pencil Method

In the wave vector k-space Fourier transform-based localization algorithm, for similar PIM source resolution or small-scale microwave components under the PIM source location

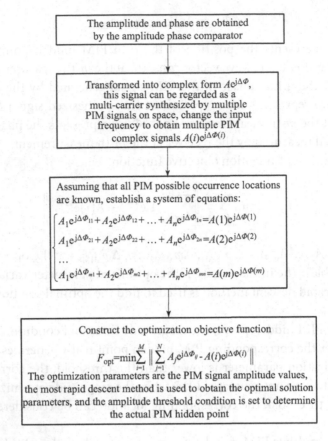

FIGURE 4.8 Flow block diagram of optimal localization algorithm for wave vector k-space inverse problem.

will be limited by the bandwidth, so in narrow-band conditions, often cannot achieve the PIM source localization. In the wave vector k-space inverse problem optimization localization algorithm, under the a priori condition that all the PIM hidden point locations are known, the optimal solution parameters are obtained by constructing the objective function with the most rapid descent method, and the magnitude threshold condition is used to determine whether each PIM hidden point actually generates PIM. However, the algorithm requires the number and location information of all possible PIM hidden points to be known. For some complex microwave components (or systems), the application of this algorithm is also limited. To address this problem, this section introduces a PIM localization algorithm based on the Matrix Pencil method, which can accurately determine the location information and magnitude of the PIM occurrence point without the support of a priori conditions (the number and location of PIM hidden points), effectively solve the localization problem of multiple PIM sources and has some robustness to noise. The specific steps are as follows:

In the first step, two coherent signals are injected into the microwave cavity to be measured under the same reference signal source, and PIM signals are generated at multiple PIM sources, respectively. The multiple PIM signals at a specific order are reflected and can be considered as a multicarrier signal after superimposed synthesis at the incident port, and the amplitude A and phase φ of the superimposed signal can be obtained through the amplitude phase comparator, which is denoted by $Aej\Phi$.

The frequency of one of the coherent signals is swept at equal intervals, where the total number of sweep points is m. For the i-th sweep, a series of amplitudes $A(i)$ and phases φ_i are obtained as

$$y(k+i\Delta k) = A(i)e^{j\varphi_i} \tag{4.21}$$

where k represents the wave vector corresponding to the lowest frequency, and Δk represents the wave vector corresponding to the sweep interval bandwidth.

In the second step, a series of single-frequency complex signals are obtained after equally spaced sweeping in the following form:

$$y(i) = y(k+i\Delta k) = A(i)e^{j\varphi_i} = x(i) + n(i) = \sum_{n=1}^{M} A_n z_n^i + n(i) \tag{4.22}$$

where $y(i)$ and $y(k+i\Delta k)$ represent the noisy signal, $x(i)$ is the signal without noise, $n(i)$ is the noise signal, M is the number of PIM sources, An is the complex amplitude, and Z_n is the pole.

The Hankel matrix Y constructed from the above noisy signals is

$$Y = \begin{bmatrix} y(0) & y(1) & \cdots & y(l) \\ y(1) & y(2) & \cdots & y(l+1) \\ \vdots & \vdots & & \vdots \\ y(m-l-1) & y(m-l) & \cdots & y(m-1) \end{bmatrix}_{(m-l)\times(l+1)} \tag{4.23}$$

where l is the Matrix Pencil parameter in the range of $[m/3, m/2]$

The singular value decomposition (SVD) is performed on the Hankel matrix Y using the following equation to obtain the singular value matrix Σ of the Hankel matrix Y:

$$Y = U\Sigma V^H \tag{4.24}$$

where U and V are the eigenvalues of YYH and YHY, respectively, and $V' = [v_1 \, v_2 ... v_M ... v_1]$ represents the matrix transpose of V.

Compare Σ with the PIM occurrence threshold to obtain the number of PIM sources that actually generate PIM in the DUT. The singular value in the matrix that is higher than the PIM occurrence threshold is considered to actually generate PIM, otherwise it is considered that no PIM is generated, and the number M of PIM sources is obtained. Among them, the PIM occurrence threshold is the ratio of the maximum value in the singular value matrix Σ.

The generalized eigenvalues of the matrix pencil $\left\{V_2'^H - \lambda V_1'^H\right\}$ are solved by the matrix pencil method so that the poles z_n can be found, where $V_1' = \begin{bmatrix} v_1 & v_2 & \cdots & v_{M-1} \end{bmatrix}$, $V_2' = \begin{bmatrix} v_2 & v_3 & \cdots & v_M \end{bmatrix}$, V_1' and V_2' are the matrices of the 1st~M-1st column and the 2nd~Mth column of V', respectively.

With the pole z_n, the following equation is used to obtain the location x_n where the PIM occurs:

$$z_n = \exp\left(j2x_n\Delta k\right) \tag{4.25}$$

By the least squares method, the magnitude of the PIM occurrence point is obtained as

$$\begin{bmatrix} y(0) \\ y(1) \\ \vdots \\ y(m-1) \end{bmatrix} = \begin{bmatrix} 1 & 1 & \cdots & 1 \\ z_1 & z_2 & \cdots & z_M \\ \vdots & \vdots & & \vdots \\ z_1^{(m-1)} & z_2^{(m-1)} & \cdots & z_M^{(m-1)} \end{bmatrix} \begin{bmatrix} A_1 \\ A_2 \\ \vdots \\ A_M \end{bmatrix} \tag{4.26}$$

where $A_1, A_2, \ldots, A_M$ are the complex amplitudes at the PIM source $x_1, x_2, \ldots, x_M$ that generate PIM, respectively.

The localization algorithm based on matrix pencil method does not need the number of possible PIM hidden points and location information to obtain the location of all possible PIM generating source points and their relative magnitude, and accurately determine the location information of PIM occurrence points, so as to effectively solve the localization problem of multiple PIM sources.

The flow block diagram of the localization algorithm based on matrix pencil method is shown in Figure 4.9.

4.4.3 Simulation Experiment Verification of PIM Localization
4.4.3.1 PIM Localization Simulation Verification

a. Verification of inverse space Fourier transform localization algorithm

In this subsection, the second harmonic generated by the diode is used to simulate the nonlinear source of the PIM, and the proposed PIM positioning method is experimentally verified. The second harmonic can be regarded as the case when the two excitation carrier signals have the same frequency, which simplifies the experimental steps without losing generality. To simulate the process of multiple PIM sources generating PIM, multiple diodes HSMS2802 are connected in parallel at different locations along the micro-strip line, and the second harmonic is generated by using the coupling energy from the micro-strip line into the diodes, which is re-injected into the main line of the micro-strip line after reflection and then superimposed at the port, as shown in Figure 4.10.

First, the circuit simulation software is used for simulation verification. Based on the diode coupling structure shown in Figure 4.10, the amplitude and phase superposition values of the second harmonic are obtained at the incident port of the

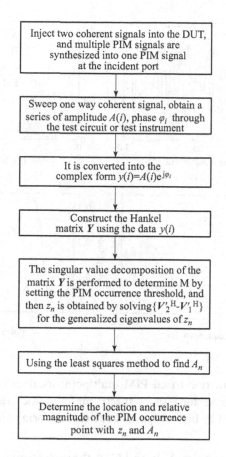

Inject two coherent signals into the DUT, and multiple PIM signals are synthesized into one PIM signal at the incident port

Sweep one way coherent signal, obtain a series of amplitude $A(i)$, phase φ_i through the test circuit or test instrument

It is converted into the complex form $y(i)=A(i)e^{j\varphi_i}$

Construct the Hankel matrix Y using the data $y(i)$

The singular value decomposition of the matrix Y is performed to determine M by setting the PIM occurrence threshold, and then z_n is obtained by solving $\{V_2'^H\text{-}V_1'^H\}$ for the generalized eigenvalues of z_n

Using the least squares method to find A_n

Determine the location and relative magnitude of the PIM occurrence point with z_n and A_n

FIGURE 4.9   Flow block diagram of the localization algorithm based on matrix pencil method.

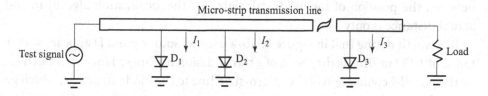

FIGURE 4.10   Schematic diagram of HSMS2802 diode coupling structure.

excitation signal source, where three diodes are placed at the positions of 1.8, 2.4 and 3 m away from the 50 Ω transmission line port, respectively. The scanning excitation source frequency is 1.125–1.275 GHz, thus obtaining a second harmonic bandwidth of 300 MHz. According to the construction method of wave vector k-space vector signal, the multicarrier signal as shown in Figure 4.11a can be obtained. It can be clearly seen that the signal in the figure is a periodic multicarrier signal.

After obtaining the wave vector k-space multicarrier signal shown in Figure 4.11a, the wave vector k-space signal spectrum of the multicarrier signal can be obtained by the Fourier transform of wave vector k-space PIM localization principle in the previous section, as shown in Figure 4.11b. It can be seen that the horizontal coordinates

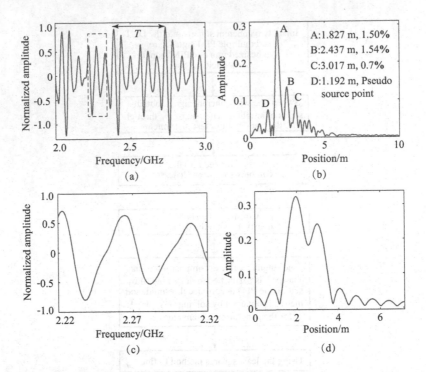

FIGURE 4.11 Software simulation-based PIM multipoint localization. (a) Wave vector k-space multicarrier signal. (b) Wave vector k-space spectrum; (c) multicarrier signal in 100 MHz bandwidth. (d) The result obtained by inverse spatial Fourier transform of (c).

of the three spectrum poles A, B and C in the spectrum are close to the position of the diode, which are 1.827, 2.437 and 3.017 m respectively, and the maximum error between the position of the diode calculated by the localization algorithm and the actual distance is only 1.54%.

It is worth noting that in Figure 4.11b, a pseudo-source point D appears with a distance of 1.192 m. This is the result of a transmission line impedance mismatch caused by the parallel connection of the micro-strip line to the diode structure, which generates reflections and thus leads to the pseudo-source point information obtained by the algorithm. In order to generate strong nonlinear harmonics, the diode requires a larger excitation signal energy, which leads to reflections of the excitation signal due to impedance mismatch. At the position of the latter diode, the excitation signal is reflected and re-injected into the former diode, where it is mixed with the source signal again to produce a second harmonic signal with a different phase difference. After a simple calculation, it can be seen that the pseudo-source point D is the result of the second harmonic mixed again with the second harmonic generated by the diode of the previous stage after the excitation signal is reflected in the latter stage. However, in the actual PIM measurement system, because the PIM signal is generally different from the excitation carrier signal by almost 100 dB or more, the existence of the PIM source does not lead to impedance mismatch of the transmission line, so there is no reflection caused by the pseudo-PIM source.

TABLE 4.1   100 MHz Optimization Results

Amplitude	Optimization Result Variable	PIM Initial Phase	Optimization Result Variable
A_1 (1.2 m)	0.077	$\Delta\varphi_{MIM1}$	1.833
A_2 (1.8 m)	0.644	$\Delta\varphi_{MIM2}$	1.043
A_3 (2.4 m)	0.341	$\Delta\varphi_{MIM3}$	1.130
A_4 (3.0 m)	0.210	$\Delta\varphi_{MIM4}$	0
A_5 (3.5 m)	0	$\Delta\varphi_{MIM5}$	0.007

b. Verification of Inverse Problem Optimization Localization Algorithm

To verify the optimized localization algorithm for the inverse problem in Section 4.3.1, here, the narrow-band case is simulated, i.e., intercepting a portion of the entire k-space multicarrier signal, as shown by the dotted box in Figure 4.11a. Figure 4.11c shows the amplified signal, which contains the second harmonic signal information in the 100 MHz bandwidth. It is directly subjected to the k-space Fourier transform, and the result shown in Figure 4.11d is obtained. Obviously, the amount of k-space signal spectrum information is insufficient due to the lack of bandwidth, which causes the overlap of actual source points and ultimately the location of the PIM source cannot be distinguished.

According to the k-space inverse problem optimization localization algorithm described in Section 4.4.1, based on prior knowledge, set all possible PIM hidden points to be 1.2, 1.8, 2.4, 3.0 and 3.5 m away from the input port of the 50 Ω transmission line, among which 1.2 and 3.5 m are artificially set redundant points to test the robustness of the algorithm. First, filter out the high-frequency clutter in the k-space spectrum; then, according to the established least square error objective function, the steepest descent method is used to optimize the 10-dimensional variable Monte Carlo. The optimization results are shown in Table 4.1. Within the error range, the amplitude optimization result at 3.5 m of the redundant point is 0, and the amplitude optimization result at 1.2 m is close to 0, which verifies that the inverse problem is used under the support of prior conditions. The optimized localization algorithm can effectively solve the localization of multiple PIM sources under narrowband conditions.

Using the optimization results to reconstruct the k-space synthesized signal, a comparison of the optimized waveform and the actual signal waveform for the inverse problem shown in Figure 4.12 is obtained. It can be seen from the figure that the optimized waveform matches well with the waveform of the actual signal within the error range.

4.4.3.2 Experimental Verification of the Localization of Diode Analog PIM Sources

a. Experimental Verification of Diode Second Harmonic Simulation PIM Source

1. Second Harmonic PIM Verification Platform Construction and Calibration

The PIM localization simulation verification (second harmonic) platform is built according to the PIM localization measurement architecture described in

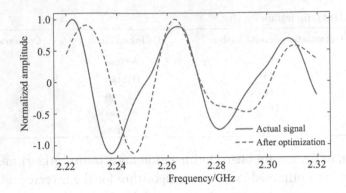

FIGURE 4.12   Comparison of inverse problem optimization waveform and actual signal waveform.

Section 4.3.2, combined with diode second harmonic signals, and its schematic block diagram is shown in Figure 4.13a. In this case, a multisegment 50 Ω coaxial cable with SMA interface is used as the transmission line, and three HSMS-2802 diodes are soldered to each 50 Ω micro-strip line as the analog PIM source. Two phase-synchronous PLL sources based on the ADF4351 are used as the excitation signal source and the "virtual" PIM reference source, respectively. The AD8302-based amplitude phase comparator is used to detect the phase and amplitude between the actual PIM signal and the "virtual" PIM reference signal. Finally, the NI USB-6251 is used as a data collector for continuous acquisition and digital storage of the phase and amplitude measurements.

Figure 4.13b shows the actual simulation verification physical platform, and two excitation signal sources are used in the experiment. The excitation source signal strength is 5 dBm. After filtering, the excitation source signal is injected into the 50 Ω coaxial cable and diode circuit through the duplexer. The lower left corner in Figure 4.13b is an enlarged view of the diode circuit, and its length is 3 cm. The harmonic signal generated by the diode and the excitation source signal are output through the duplexer, and then the excitation signal is filtered through a filter to obtain a relatively pure second harmonic signal, which is input to the amplitude phase comparator. The other end of the amplitude and phase comparator AD8302 is connected to the "virtual" PIM reference signal attenuated by the filter and attenuator. Using software to control the automatic scanning frequency, the frequency of the excitation source signal is scanned from 1.125 to 1.275 GHz with the stepped frequency of 0.2 MHz, and the corresponding "virtual" PIM reference signal frequency is scanned from 2.25 to 2.55 GHz with the stepped frequency of 0.4 MHz.

Before the actual localization measurement, the entire system needs to be phase-calibrated, that is, to remove the phase offset caused by devices other than the PIM source. Its specific operation is that a 1 m long coaxial cable is connected to a diode circuit (in this example, the actual distance between the incident port and the diode is 1.015 m), the automatic frequency sweep is performed under the software control to make the excitation signal source frequency scan from 1.125 to

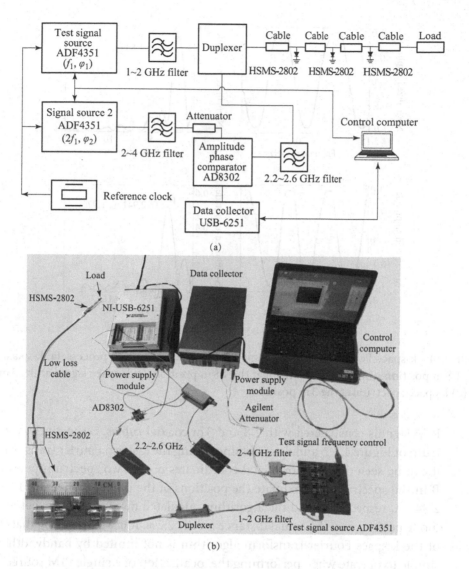

FIGURE 4.13 PIM localization simulation verification (second harmonic) platform. (a) Schematic block diagram. (b) Physical platform.

1.275 GHz with the stepped frequency of 0.2 MHz, and the amplitude and phase difference between the second harmonic and the "virtual" PIM reference source are measured through the amplitude and phase comparator AD8302. Obviously, the phase shift of the transmission line and the diode circuit used for calibration can be obtained by measurement, so the phase shift caused by the filter, coupler, attenuator and other system components can be calibrated.

2. Verification of single point PIM source localization

First, the localization of a single PIM source is verified. After system calibration, different lengths of coaxial cables are used to simulate PIM sources at different locations. Figure 4.14a and c show the vector signal forms of the simulated

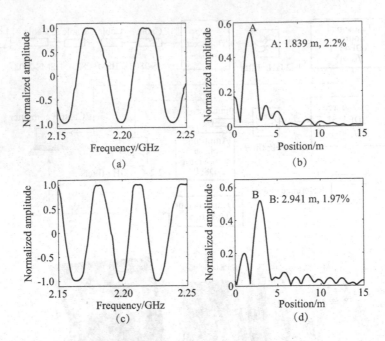

FIGURE 4.14   k-space Fourier transform single PIM source localization process. (a) k-space signal at the 1.8 m position. (b) k-space spectrum at the 1.8 m position. (c) k-space signal at the 3 m position. (d) k-space spectrum at the 3 m position.

PIM signals constructed with 1.8 and 3 m coaxial cables, respectively; the spectrum of Figure 4.14b and d are obtained using the k-space Fourier transform, and it can be seen that the horizontal coordinates of the two spectrum poles A and B in the spectrum approximate the positions of the diodes, which are 1.859 and 2.941 m, respectively, and the algorithm calculated maximum error between the diode position and the actual distance is only 2.2%. It is worth noting that the use of the k-space Fourier transform algorithm is not limited by bandwidth and is simple to operate when performing the localization of a single PIM source.

3. Multipoint PIM source localization verification

The multipoint PIM source localization verification is similar to the simulation structure of the single-point PIM source localization verification, which uses three sections of coaxial cable lines with lengths of 1.8, 0.6 and 0.6 m to connect three diode circuits, and accordingly, the positions of the three diodes are 1.815, 2.43 and 3.06 m from the incident port (taking into account the distance error of the actual diode welding installation). The excitation source frequency was scanned from 1.125 to 1.275 GHz with the stepped frequency of 10 MHz, and a second harmonic bandwidth of 300 MHz was obtained. Similarly, the k-space Fourier transform localization algorithm was used to obtain the k-space multicarrier vector signal form and the Fourier transformed k-space spectrum, as shown in Figure 4.15a and b. It can be seen that the diode positions are precisely located at 1.835, 2.474, and 3.010 m, with errors of 1.1%~1.7%. At the same time,

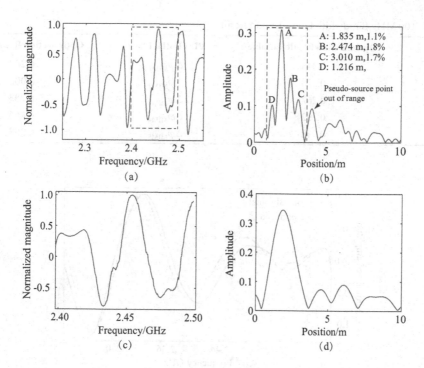

FIGURE 4.15 k-space Fourier transform multipoint PIM source localization process. (a) Multicarrier signals in the full bandwidth of k-space. (b) Spectrum in k-space. (c) Multicarrier signal within 100 MHz in k-space. (d) Spectrum after k-space Fourier transform of (c).

the pseudo-source point due to diode energy absorption and impedance mismatch is also located at the same time with a distance of 1.216 m, which corresponds to the simulation results. It is worth noting that since the diode circuit size is very small, the wave vector of the micro-strip line in the diode circuit is assumed to be close to the wave vector of the coaxial cable, thus approximating a uniform transmission line, and this approximation is to facilitate the length calculation. When the length of the micro-strip line is comparable to that of the coaxial cable line, it is necessary to consider both the wave vector of the micro-strip line and the wave vector characteristics of the coaxial cable line, and thus the phase within the nonuniform cavity needs to be used to measure and calculate the correspondence between phase and distance.

Similarly, the sampling bandwidth of the second harmonic can be reduced to 100 MHz, and the k-space spectrum obtained by the k-space Fourier transform localization algorithm, but the spectral resolution is limited, causing the source point position to overlap, so localization cannot be achieved.

According to the k-space inverse problem optimization localization method, a priori knowledge is first obtained, and all possible source locations are preset to be 1.2, 1.815, 2.43, 3.06 and 3.5 m from the 50 Ω transmission line input port, where the redundant point at 3.5 m is used to check the robustness of the algorithm. After filtering, Monte Carlo optimization of the 10-dimensional variables is performed using the fastest descent method. 100 MHz bandwidth optimization

TABLE 4.2   100 MHz Bandwidth Optimization Results

Amplitude	Optimization Result Variable	PIM Initial Phase	Optimization Result Variable
A_1 (1.2 m)	0.346	$\Delta\varphi_{MIM1}$	6.283
A_2 (1.815 m)	0.728	$\Delta\varphi_{MIM2}$	3.008
A_3 (2.43 m)	0.267	$\Delta\varphi_{MIM3}$	6.283
A_4 (3.06 m)	0.292	$\Delta\varphi_{MIM4}$	4.866
A_5 (3.5 m)	0.098	$\Delta\varphi_{MIM5}$	6.283

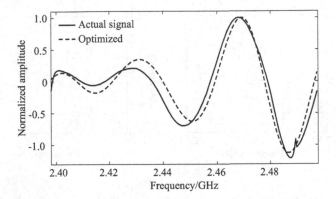

FIGURE 4.16   Comparison of optimized waveform and actual waveform.

results are shown in Table 4.2. Within the error range, the magnitude optimization results at the redundant point 3.5 m differed from the other points by one order of magnitude. The waveform comparison shown in Figure 4.16 is obtained by reconstructing the k-space multicarrier signal using the optimization results, and the optimized waveform matches well with the waveform of the actual signal within the error range.

b. Experimental Verification of Diode 3rd-order Intermodulation Simulated PIM Source Localization

1. The construction and calibration of Diode 3rd-order intermodulation simulation PIM localization verification system

Figure 4.17a shows the block diagram of the PIM localization verification based on diode 3rd order intermodulation simulation. Among them, the SMA interface with multiple segments of 50 Ω coaxial cable is used as the transmission line, and three HSMS-2802 diodes are soldered to the 50 Ω micro-strip line as the simulated PIM sources. Three ADF4351-based phase-synchronous PLL sources are used as two-tone excitation signal sources and "virtual" PIM reference sources, respectively. A broadband oscilloscope (R&T RTE 1104 5GHz) is used instead of AD8302 to acquire the phase and amplitude of the actual PIM signal and the "virtual" PIM reference signal. Finally, a computer is used to process the acquired data to obtain the required amplitude and phase information.

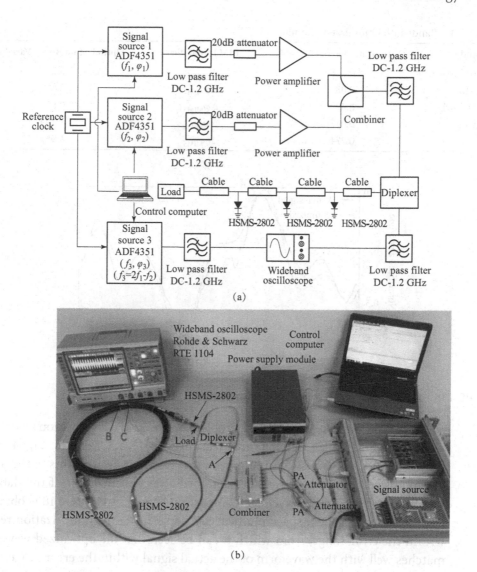

FIGURE 4.17 PIM positioning simulation verification (3rd-order intermodulation) platform. (a) Principle block diagram; (b) physical platform.

Figure 4.17b shows the actual simulated validation physical platform with a 3-way coherent excitation signal source used in the experiment. The lengths of the three coaxial lines in the circuit are 0.6, 1.0 and 3.0 m, so the simulated PIM source locations are 0.615, 1.665 and 4.825 m (taking into account the distance error of the actual diode welding installation). One of the two coherent sources has a fixed frequency of 950 MHz, and the other has a frequency that scans from 899.5 to 1000.5 MHz with the step of 1 MHz. The corresponding "virtual" PIM reference signal scans from 1000.5 to 899.5 MHz with the step of 1 MHz.

2. Localization verification of k-space inverse problem optimization multipoint PIM source

TABLE 4.3   Bandwidth Optimization Results

Amplitude	Optimization Result Variable	PIM Initial Phase	Optimization Result Variable/rad
A_1 (0.615 m)	0.438	$\Delta\varphi_{MIM1}$	3.392
A_2 (1.3 m)	0.082	$\Delta\varphi_{MIM2}$	2.099
A_3 (1.665 m)	0.379	$\Delta\varphi_{MIM3}$	0.627
A_4 (3.5 m)	0.047	$\Delta\varphi_{MIM4}$	1.744
A_5 (4.825 m)	0.274	$\Delta\varphi_{MIM5}$	1.952

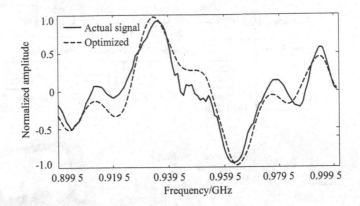

FIGURE 4.18   Optimized waveform vs. actual waveform.

According to the k-space inverse problem optimization localization algorithm, all possible PIM source locations are preset to be 0.615, 1.3, 1.665, 3.5 and 4.825 m from the 50 Ω transmission line input port, respectively, and this is the a priori condition obtained. The results obtained after optimization of the data are shown in Table 4.3. The waveform comparison shown in Figure 4.18 is obtained by reconstructing the k-space multicarrier signal using the optimization results within the error range, from which it can be seen that the optimized waveform matches well with the waveform of the actual signal within the error range.

3. PIM multipoint localization by Matrix Pencil method

In order to apply the Matrix Pencil method to achieve PIM localization, it is necessary to calibrate the calculated data, so the localization distance is

$$x_i = \arg(z_i - z_{cal})/(2\Delta k),$$

where z_i represents the theoretical pole of the i-th PIM source; z_{cal} represents the calibration of the theoretical pole; $\arg z_{cal} = 0.05$ rad, $\Delta k = 0.027$ rad/m.

Without setting any a priori conditions, the data obtained from sampling were processed and optimized using the matrix pencil method, and the results obtained are shown in Table 4.4. As can be seen from the table, when $N=3$, the accurate localization of three PIM sources is obtained, and the maximum localization error is 6.7%, and the matrix pencil method can effectively achieve the accurate localization of multiple PIM sources.

TABLE 4.4   100 MHz Bandwidth Matrix Beam Optimization Results

N	I	$\arg_{zi}/\text{rad}$	x_i/m	localization error
1	1	0.102	0.963	—
2	1	0.103	0.981	—
	2	0.312	4.852	—
3	1	0.081	0.574	6.7
	2	0.143	1.722	3.4
	3	0.313	4.870	0.9
4	1	0.081	0.574	6.7
	2	0.144	1.741	4.6
	3	0.313	4.870	0.9
	4	0.453	7.463	—

BIBLIOGRAPHY

1. Zhang M., Zheng C., Wang X., et al. Localization of PIM based on the concept of k-space multicarrier signal. *IEEE Transactions on Microwave Theory and Techniques*, 2017, 65(12): 4997–5008.
2. Ye M., He Y., Wang X., et al. nonlinear physical mechanism and calculation method of passive intermodulation at metal waveguide connection. *Journal of Xi'An Jiaotong University*, 2011, 45(2): 82–86.
3. Li J., Zhao X., Gao F., et al. Passive intermodulation of metal point contact structure based on electrothermal coupling effect. *Journal of Xi'An Jiaotong University*, 2018, 52(9): 76–81.
4. He Y., Li J., Cui W., et al. Review of research on distributed passive intermodulation. *Space Electronic Technology*, 2016, 13(4): 1–6.
5. Gao F., Zhao X., Ye M., et al. A passive intermodulation measuring method based on the coupling of dipole near-field. *Space Electronic Technology*, 2018, 15 (3): 15–21.
6. Bennett J. C., Anderson A. P., Mcinnes P., et al. Microwave holographic metrology of large reflector antennas. *IEEE Transactions on Antennas and Propagation*, 1976, 24 (3): 295–303.
7. Hienonen S., Raisanen A. V. PIM near – field measurements on micro-strip lines. *The 34th European Microwave Conference*, 2004, 1041–1044.
8. Mantovani J. C., Denny H. W., Warren W. B. Apparatus for Locating PIM Interference Sources. United States, 1987.
9. Kim J. T., Cho I. K., Jeong M. Y., et al. Effects of external PIM sources on antenna PIM measurements. *Etri Journal*, 2002, 24 (6): 435–442.
10. Aspden P. L., Anderson A. P., Bennett J. C. Microwave holographic imaging of intermodulation product sources applied to reflector antennas. *International Conference on Antennas and Propagation*, 1989, 463–467.
11. Yang S., Wu W., Xu S., et al. A PIM source identification measurement system using a vibration modulation method. *IEEE Transactions on Electromagnetic Compatibility*, 2017, 59 (6): 1677–1684.
12. Chen Z., Zhang Y., Dong S., et al. Wideband architecture for PIM localization. *IEEE MTT – S International Wireless Symposium (IWS)*, Chengdu, China, 2018: 1–6.
13. Aspden P. L., Anderson A. P. Identification of PIM product generation on microwave reflecting surfaces. *IEEE Proceedings H Microwaves, Antennas and Propagation*, 2002, 139 (4) : 337–342.
14. Dahele J. S., Cullen A. L. electric probe measurements on micro-strip. *IEEE Transactions On Microwave Theory and Techniques*, 1980, 28 (7): 752–755.
15. Shitvov A. P., Zelenchuk D. E., Schuchinsky A., et al. PIM generation on printed lines: Near – field probing and observations. *IEEE Transactions on Microwave Theory and Techniques*, 2008, 56 (12): 3121–3128.

16. Hienonen S., Golikov V., Vainikainen P., et al. Near – field scanner for the detection of pim sources in base station antennas. *IEEE Transactions on Electromagnetic Compatibility*, 2004, 46 (4): 661–667.
17. Hienmnen S., Golikov V., Mottonen V.S., et al. Near – field amplitude measurement of PIM in antennas. *Proceedings of the 31st European Microwave Conference*, London, 2001, 1–4.
18. Hienonen S., Vainikainen P., Raisanen A. V. Sensitivity measurements of a PIM near – field scanner. *IEEE Antennas & Propagation Society*, 2003, 45 (4): 124–129.
19. Walker A., Steer M., Gard K. Simple broadband relative phase measurement of intermodulation products. *65th ARFTG Conference Digest*, 2005, 123–127.
20. Walker A., Steer M., Gard K. A vector intermodulation analyzer applied to behavioral modeling of nonlinear amplifiers with memory. *IEEE Transactions on Microwave Theory and Techniques*, 2006, 54 (5): 1991–1999.
21. Sarkar T. K., Pereira O. Using the matrix pencil method to estimate the parameters of a sum of complex exponentials. *IEEE Antennas & Propagation Society*, 1995, 37 (1): 48–55.

CHAPTER 5

Passive Intermodulation Detection Technology for Microwave Components

Xiang Chen, Jun Li, Wanzhao Cui, and Xinbo Wang

CONTENTS

DOI: 10.1201/9781003269953-5

5.1 OVERVIEW

The mechanism study of passive intermodulation (PIM), the accuracy verification of the analysis and prediction results and the validation of the effectiveness of the suppression technique are inseparable from the accurate detection technology, so the high-performance detection technology is a key measure of the accurate PIM performance of microwave components. For microwave components applied in spacecraft, the requirements for PIM inspection are higher due to the special nature of their working environment. The physical complexity of PIM is diverse and unpredictable in time, power level, temperature, stress, aging, etc. In addition, temperature variations in the satellite-based environment lead to large and frequent changes in the temperature and stress properties of PIM materials. For satellite-based applications, there are several key issues for the accurate detection of PIM as follows:

1. High-power transmit carrier and low-power PIM signal. Because the carrier power is relatively large, and the PIM signal is extremely weak, the dynamic range in the satellite high-power system of about 200 dBc is achieved generally using transceiver duplexer and receive filter. For low-order PIM (especially the 3rd order), whose frequency point is close to the transmitting carrier, it is not easy to achieve transceiver isolation and out-of-band suppression.

2. Ultra-low residual PIM requirements for the test system. In order to achieve high detection sensitivity, the residual PIM level of the test system itself needs to be maintained at a very low level.

3. Detection accuracy and real-time. PIM signal often shows random jumping characteristics. By using traditional spectrometer for detection equipment, the bottom noise of the spectrometer must be reduced as much as possible to measure the weak PIM signal (−150 dBm), resulting in a slow scan speed, which may lead to missed detection, and therefore cannot meet the detection of real-time requirements.

In addition to the above problems, there are also problems in high-power detection, simultaneous detection of multiple physical environment parameters and multicarrier detection. Conventional detection methods are difficult to meet these high-performance detection needs.

In summary, it is necessary to meet simultaneously the basic requirements of high sensitivity, wide band coverage, temperature change test and high-power test to achieve high performance PIM detection of high-power microwave components in spacecraft under the consideration of the simultaneous detection of multiple physical parameters, multicarrier detection and other issues for the deeper needs. The real-time detection is also put forward for high requirements of the special characteristics of microwave components for spacecraft. To meet these requirements, in which many actual detection requirements are contradictory to each other, it is necessary to carry out the overall top-level design of detection methods. Therefore, how to propose innovative methods to resolve technical contradictions in order to achieve multifunctional, multidemand synthesis is one of the difficulties in achieving high-performance testing. In addition, the low PIM performance of the test system itself is the key to ensure that it achieves high sensitivity detection, in order to meet a variety of needs at the same time; how to achieve a stable low PIM performance of the system itself is also an important difficulty, which puts forward very high requirements for the design of the key core components of the system. For real-time detection needs, conventional detection equipment for PIM small signal detection speed is too slow to meet the detection of real-time, so how to achieve rapid real-time detection of weak PIM signal is also a key challenge, which involves high-sensitivity signal detection and real-time signal processing technology.

5.2 ADVANCES IN PASSIVE INTERMODULATION DETECTION TECHNOLOGY

Since the mechanism of PIM generation is very complex, there is no perfect theoretical model and effective analysis and evaluation means, and it mainly relies on experimental tests to measure the PIM level of components or subsystems in engineering design, so the high-performance PIM detection technology is very critical. Attention has been paid to the detection of PIM at home and abroad, and there are a series of measurement instruments, basically to meet the needs of PIM detection in civil communication systems. The general test methods for PIM are more mature, and the International Electrotechnical Commission (IEC) has proposed two recommended test methods for testing PIM signals, namely the reflection method and transmission method. The basic principle of these two test methods is as follows: first, the system provides two equal-level signals, and then the two signals are combined into one signal, which is fed back to the component under test, and the PIM level generated by the component under test is required to be about 10 dB higher than the measurement system's own PIM level; then, the n-order PIM product is detected by the receiving device. The basic block diagram of the reflective and transmission methods of PIM measurement is shown in Figure 5.1. In comparison, the reflection method measurement is simple to implement; the transmission method measurement is more comprehensive and can measure the worst PIM state of the microwave components.

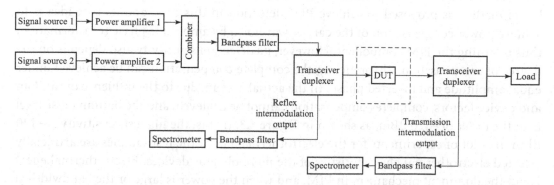

FIGURE 5.1 Schematic diagram of a generic PIM test system.

With the increasing demand, the main requirements for PIM detection methods focus on the detection sensitivity and detection speed, in which the detection sensitivity mainly depends on the test system's own PIM level, and the detection speed mainly depends on the measurement speed of the end signal receiving detection equipment. In order to study the characteristics of PIM level changes with time as well as environmental conditions, loading long-term real-time measurements under different environmental conditions is very important for PIM detection.

In addition, wider test bandwidth, larger test power range, PIM testing under multicarrier conditions and modulated signal PIM testing are also the focus of future PIM test system development.

Radiative (such as antennas) PIM test method is essentially the same as the standard PIM test method; however, due to their different radiation characteristics, radiative PIM measurement differs from the PIM measurement of closed cavity structures in the specific form of implementation. First, in order to avoid the influence of external interference, open structure PIM test needs to be conducted in the wave-absorbing darkroom, whose PIM level must be maintained at a very low level (usually lower than the detection sensitivity) to ensure test accuracy and test consistency. Open structure PIM measurement is also divided into reflective and radiative, in which the reflective method is similar to the standard PIM measurement method, only the placement environment of the DUT is different; in radiative method, radiation field can replace the transceiver duplexer in the standard method, and the transmitting antenna directly radiates the signal into the open space; the PIM signal of interest is detected and received by the receiving antenna placed at a specific location. In the radiative measurement, the transmit signal can either be radiated separately with multiple antennas or synthesized and radiated through one antenna, which should be selected according to different needs.

In 2010, scholars of the North Carolina State University conducted a systematic study of the electrothermal coupling PIM effect and proposed a test scheme for electrothermal coupling PIM testing, as shown in Figure 5.2. The PIM induced by the electrothermal coupling effect is only effective under the condition that the carrier frequency interval is very narrow (kHz level), and when measured by the conventional PIM test system, the low-order PIM products will be very close to the transmitting frequency at such a narrow frequency interval and cannot be separated by filtering. For this reason, a power-compensation-based method is proposed to achieve PIM detection in this case, and its core idea is to achieve power compensation of the carrier waveform by amplitude phase transformation, thus retaining the PIM component. However, its detection sensitivity also depends on the degree of compensation. Theoretically, the complete compensation can be achieved under equal amplitude and inverted phase. In the actual system, due to the influence of random and device factors, complete compensation cannot be achieved, and the bottom noise level is in the order of −120 dBm, as shown in Figure 5.3, that is, the highest sensitivity is −120 dBm. In a lot of experiments for the electrothermal effect, the test samples are artificially selected electrothermal characteristics of the more obvious devices. Electrothermal effect is not the dominant mechanism in PIM, and when the power is large or the bandwidth is wide, the limitation of this power compensation method is very obvious. Therefore, the

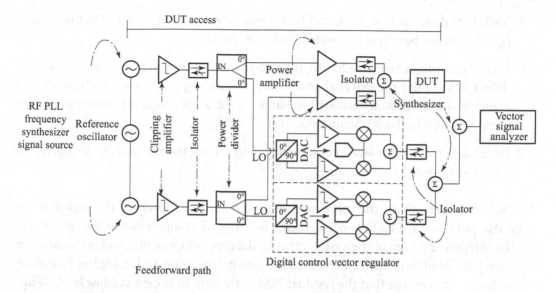

FIGURE 5.2 PIM detection method based on compensation idea.

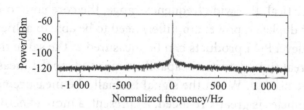

FIGURE 5.3 Bottom noise level of the PIM test system by compensation method.

PIM test system designed on this basis is not representative except for its compensation idea.

In the field of high-power microwave technology, the research institutions, such as the European Space Agency (ESA) and the State Key Laboratory of Space Microwave Technology, have been conducting related research and development capacity building for a long time, and more high-performance PIM test systems have been established for different application needs. The main development of PIM detection tends to the following aspects:

1. High sensitivity – to detect much weaker PIM signals.

2. Wide frequency bandwidth – to simultaneously measure multiorder PIM as well as PIM at multiple frequency intervals.

3. High power detection – to achieve PIM measurement with continuously adjustable carrier power over a wide range.

4. Real-time long time detection – to capture and record the transient mutation signal, and to realize long-time high capacity test.

5. Multifunction, multichannel, and multiband detection – to detect PIM under multiple frequency bands, multicarrier conditions and modulated carrier.

6. Loading measurement under high-performance environmental conditions – to detect PIM under the external environmental loading conditions, such as a wider range of the temperature and/or mechanics, for example, high and low temperature PIM detection for large antenna systems.

7. Integrated fully automated measurement – to realize the integrated intelligence of PIM test system.

The higher the sensitivity, the more accurate the detection. However, the higher sensitivity also puts forward higher requirements for the core components of the test system and the integrated design of the system, that is, the core components (such as transceiver duplexer, filter, load) and the connectors in the system have good and stable low PIM characteristics, so as to ensure that the residual PIM of the system is kept at a low level. When the sensitivity is very high, the PIM index of the components in the test system is very strict, so the implementation of high-power microwave passive components with ultra-low PIM becomes critical. For wide frequency range, the core components of the system (such as transceiver duplexer, power amplifier) need to be able to achieve broadband coverage, so that multilevel PIM products can be measured at the same time to improve the test efficiency. For real-time detection capability, more advanced measurement methods and instruments are needed. When the signal is small, the measurement speed of traditional swept spectrometer is greatly reduced. At present, a more effective method is to use real-time signal analyzer or vector signal analyzer, with which the detection speed can be improved greatly, and the real-time measurement can be realized basically under the condition of the same accuracy. Large power range is conducive to the study of the PIM characteristics of the device under different power inputs. Multiband PIM test mainly relies on core equipment such as amplifiers, and the essence of the test method is not special. In summary, it can be seen that the PIM test system is a comprehensive test application system with complex system functional components and high development and construction costs.

5.3 BROADBAND HIGH-SENSITIVITY PASSIVE INTERMODULATION DETECTION METHOD AND ITS IMPLEMENTATION

Most of the existing PIM test systems are designed for a single PIM test requirement, such as the 3rd order PIM test usually designed for two-way 20 W test power. As shown in Figure 5.1, when the combiner uses a synthetic duplexer, it can only meet the PIM test under a specific narrow bandwidth carrier and cannot realize the broadband sweep and variable frequency interval PIM test. However, if the 3 dB bridge synthesis is used to simply meet the broadband synthesis requirements, it will cause the system to generate an additional 3 dB of power loss, so that the test power range of the DUT is limited.

In order to achieve wider receiver band coverage and more-order PIM testing at the same time, the transceiver duplexer is required to have a wider receiver bandwidth; in

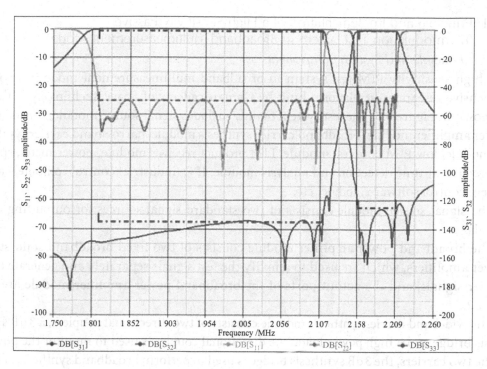

FIGURE 5.4 Transceiver duplexer simulation results.

order to ensure that the system has a high enough sensitivity, the transceiver duplexer is required to have better low PIM characteristics and higher transceiver isolation. For example, if the two carriers are separated by 50 MHz and the 3rd to 15th order PIM is measured simultaneously, the Tx bandwidth of the transceiver duplexer is required to be 50 MHz, the Rx bandwidth should be 300 MHz, the Tx passband and Rx passband should be separated by 50 MHz, and the isolation is usually required to be at least 100 dB. According to the simulation results shown in Figure 5.4, it is difficult to achieve wide receiving bandwidth and high isolation at the same time. To achieve them simultaneously, the design of Tx channel and Rx filter will become very complicated and need to meet the low PIM characteristics at the same time, but this has a high risk of implementation and cannot meet the coverage in the broadband range, so that only a few orders PIMs can be tested, and it is difficult to measure more-order PIM signals at the same time.

The above problems make it difficult to form a comprehensive and efficient PIM test system. If PIM testing is conducted under multiple conditions, additional test systems need to be temporarily built and calibrated, making the test complex and inefficient. In addition, the low PIM test system needs to maintain a stable connection status, and frequent disassembly of the connection is not conducive to the low PIM characteristics of the system, which will affect the sensitivity of the test.

Taking into account the development needs and the progress made in the work, this section gives the principles and implementation process of two high-performance PIM inspection methods to address these problems in the current PIM inspection technology.

5.3.1 Principle and Implementation of a High-Sensitivity Passive Intermodulation Test System in Broadband Multimeasurement Mode

5.3.1.1 Principle and System Components

The high-sensitivity PIM test system in broadband multimeasurement mode is a comprehensive research test platform built by China Academy of Space Technology (Xi'an), as shown in Figure 5.5. The test system mainly includes signal source module, broadband power amplifier module, broadband carrier synthesis module, narrowband carrier synthesis module, low PIM core test module, PIM product receiver module, power meter, spectrometer, real-time detection and analysis module, IPC (industrial control computer) and system control software module, etc.

The signal source module consists of two standard signal sources for outputting test signals.

The broadband power amplifier module consists of two fixed broadband solid-state power amplifiers, which are used to amplify the test signals separately and generate two carrier signals, and the output ports of the broadband power amplifier module are P1 and P2.

The wideband carrier synthesis module consists of two directional couplers, a 3 dB synthesis bridge and a high power load. The directional coupler is used to monitor the power of the two carriers, the 3 dB synthesis bridge is used to perform broadband synthesis of the two carriers, and the high power load is used to absorb the synthesis loss power. The input ports of the broadband carrier synthesis module are P11 and P12, which are internally connected to the inputs of the two directional couplers, respectively; the output port is POUT1, which is internally connected to the power synthesis output of the 3 dB synthesis bridge; the power monitoring ports are P_C11 and P_C12, which are externally connected to the input ports P_C1 and P_C2 of the power meter, respectively.

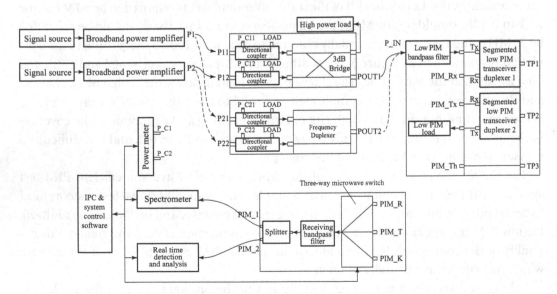

FIGURE 5.5 Principle of the high sensitivity PIM test system in broadband multimeasurement mode.

The narrowband carrier synthesis module consists of two directional couplers and a synthesized duplexer. The directional coupler is used to monitor the power of the two carriers, and the synthesizer duplexer is used to perform narrowband synthesis of the two carriers. The input ports of the narrowband carrier synthesis module are P21 and P22, which are internally connected to the inputs of the two directional couplers, respectively; the output port is POUT2, which is internally connected to the output of the synthesized duplexer; the power monitoring ports are P_C21 and P_C22, which are externally connected to the input ports P_C1 and P_C2 of the power meter, respectively.

The broadband carrier synthesis module and the narrowband carrier synthesis module cannot work simultaneously.

The low-PIM core test module includes a low-PIM bandpass filter, segmented low-PIM transceiver duplexer 1, segmented low-PIM transceiver duplexer 2 and low-PIM load. The low PIM bandpass filter further filters the synthesized carrier and suppresses the PIM products generated by the front-end components. Segmented low PIM transceiver duplexer 1 is used to separate the reflected PIM products generated by the DUT, segmented low PIM transceiver duplexer 2 is used to separate the transmitted PIM products generated by the DUT, and low PIM load is used to absorb the carrier wave. The low PIM core test module has one input port P_IN, three test ports TP1, TP2 and TP3, and three output ports PIM_Rx, PIM_Tx and PIM_Th. Inside the module, P_IN is connected to the input of the low PIM bandpass filter, and TP1 and PIM_Rx are connected to the common and Rx terminals of the segmented low PIM transceiver duplexer 1, respectively. TP2 and PIM_Tx are connected to the common terminal and Rx terminal of segmented low PIM transceiver duplexer 2, respectively. TP3 and PIM_Th are independent expansion ports with internal straight-through connections. In the low PIM core test module, the performance indexes of segmented low PIM transceiver duplexer 1 and segmented low PIM transceiver duplexer 2 are the same, but the functions are different. When different bandwidth PIM tests are required, the segmented low PIM transceiver duplexer 1 and segmented low PIM transceiver duplexer 2 of the corresponding receive band need to be replaced at the same time.

PIM product receiving module includes three-way microwave switch, receiving bandpass filter, and splitter. The three-way microwave switch completes the switching of reflected PIM signal, transmitted PIM signal and expanded port PIM signal; the receiving band-pass filter further filters the PIM signal to suppress the carrier and other interference signals outside the receiving band; the splitter divides the PIM signal into two channels. PIM product receiving module has three input ports, namely PIM_ R, PIM_ T and PIM_ K. The internal part is connected with three inputs of three microwave switches, and the external part is connected with three output ports of low PIM core test module_ Rx, PIM_ TX and PIM_ Th connection. The PIM product receiver module has two output ports, namely PIM_1 and PIM_2, which are internally connected to the two outputs of the splitter and externally connected to the spectrometer and the real-time detection and analysis module, respectively.

The power meter monitors the forward power of the two carriers in real time. The two input ports P_C1 and P_C2 are externally connected to the power monitoring ports of the wideband carrier synthesis module or the narrowband carrier synthesis module, respectively.

The spectrometer is a standard instrument, and the real-time detection and analysis module is a special custom-developed instrument, which is described in detail in Section 5.4.

The IPC and the system control software module communicate through the LAN port to complete the access control of each part of the module and realize the functions of the system such as switch operation, error calibration, data acquisition and record storage.

5.3.1.2 Functional Features

In this scheme, the mode of switching between broadband carrier synthesis and narrowband carrier synthesis is proposed to take into account the requirements.

When the carrier frequency interval should be changed in a wider frequency band or the carrier needs to be swept for testing, the wideband carrier synthesis module is selected, and the output ports P1 and P2 of the wideband power amplifier module are connected to the input ports P11 and P12 of the wideband carrier synthesis module, respectively, and the output port POUT1 of the wideband carrier synthesis module is connected to the input port P_IN of the low-PIM core test module. The coupling ports P_C11 and P_C12 of the directional coupler in the wideband carrier synthesis module are connected to the input ports P_C1 and P_C2 of the power meter for monitoring and calibrating the carrier power, respectively. In this mode, carrier powers can be synthesized at arbitrary frequency intervals over a very wide bandwidth of the bridge.

For PIM testing at a fixed carrier frequency, the narrowband carrier synthesis module can be selected by connecting the output ports P1 and P2 of the broadband power amplifier module to the input ports P21 and P22 of the narrowband carrier synthesis module, respectively, and the output port POUT2 of the narrowband carrier synthesis module to the input port P_IN of the low PIM core test module. The coupling ports P_C21 and P_C22 of the directional coupler in the narrowband carrier synthesis module are connected to the input ports P_C1 and P_C2 of the power meter, respectively, for monitoring and calibrating the carrier power. In this mode, maximum synthesized power can be ensured and additional power loss can be avoided, which facilitates PIM testing of the device port to be tested over a wide power range.

Through this switching method, both the conventional fixed carrier frequency under the high-power range of PIM test and the carrier sweep and variable frequency interval under the PIM test can be achieved.

In addition, in this solution, a wider receiving bandwidth can be achieved through the segmented low PIM transceiver duplexer mode, so that more PIM products can be measured simultaneously.

When testing different orders of PIM products, either segmented low PIM transceiver duplexer 1 or segmented low PIM transceiver duplexer 2 within the corresponding receive bandwidth can be selected, and flexible configurations can also be made by selecting segmented low PIM transceiver duplexers with different receive bandwidths to achieve the expansion of the receive bandwidth, so as to cover a wider receive bandwidth and test more orders of PIM products, as shown in Figure 5.6.

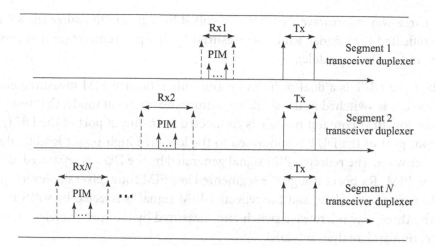

FIGURE 5.6 Schematic diagram of extended reception bandwidth of segmented low PIM transceiver duplexer.

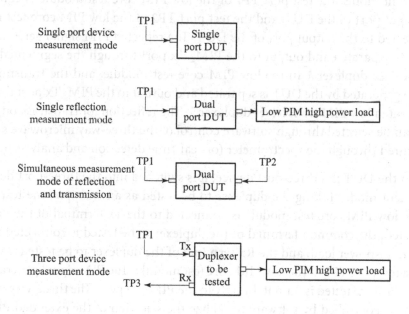

FIGURE 5.7 Schematic diagram of measurement mode.

This test system can simultaneously measure the single-port device reflective PIM measurement, two-port device reflection and transmissive PIM measurement and three-port device extended PIM, as shown in Figure 5.7.

1. When the DUT is a single-port device (e.g., load), it is switched in the single-port device measurement mode, and only reflective PIM measurements are performed. Connect the test port TP1 of the low PIM core test module to the DUT. The reflected PIM signal generated by the DUT is separated and output to the PIM_Rx port through the segmented low PIM transceiver duplexer 1 in the low PIM core test module, and

the three-way microwave switch is controlled by software to realize the selection of the reflected PIM signal, which is measured by the spectrometer (or real-time detection and analysis module).

2. When the DUT is a dual-port device and only reflective PIM measurement is performed, it is switched in the single reflection measurement mode, that test port TP1 of the low PIM core test module is connected to the input port of the DUT, and the output port of the DUT is connected to the low PIM high-power load to absorb the carrier wave. The reflected PIM signal generated by the DUT is separated and output to the PIM_Rx port through the segmented low PIM transceiver duplexer1 in the low PIM core test module, and the reflected PIM signal is selected by software control of the three-way microwave switch and measured by the spectrometer (or real-time detection and analysis module).

3. When the DUT is a dual-port device and both reflection and transmission measurements are required, it is switched into the reflection, transmission simultaneous measurement mode. The test port TP1 of the low PIM core test module is connected to the input port of the DUT, and the test port TP2 of the low PIM core test module is connected to the output port of the DUT. The reflected PIM signal generated by the DUT is separated and output to the PIM_Rx port through the segmented low PIM transceiver duplexer 1 in the low PIM core test module, and the transmitted PIM signal generated by the DUT is separated and output to the PIM_Tx port through the segmented low PIM transceiver duplexer 2. The reflection (or transmission) PIM signal can be selected through software control of the three-way microwave switch and measured through the spectrometer (or real-time detection and analysis module).

4. When the DUT is a three-port device, it is switched into the three-port device measurement mode. Taking the duplexer to be tested as an example, the test port TP1 of the low PIM core test module is connected to the Tx terminal of the duplexer to be tested, the common terminal of the duplexer to be tested is connected to the low PIM high-power load, and the Rx terminal of the duplexer to be tested is connected to the test port TP3 of the low PIM core test module. The PIM signal generated by the duplexer to be tested is output directly to the PIM_Th port. The three-way microwave switch is controlled by software to realize the selection of the extended PIM signal, and the measurement is performed by the spectrometer (or real-time detection and analysis module).

5.3.1.3 System Calibration Method

The power calibration of the system is divided into transmitting carrier signal power calibration and receiving PIM signal power calibration, whose main function is to calibrate the loss on the signal transmission link, so as to ensure the accuracy of the test results. Next, the broadband carrier synthesis mode is used as an example for illustration.

Transmitting carrier signal power calibration is to ensure real-time control of the test power at the DUT during the test process, while requiring the power measured by the power meter to be the actual power input to the DUT, that is, to calibrate the sum of all losses on the link from the power coupling ports P_C11 and P_C12 of the directional coupler in the broadband carrier synthesis module to the test port TP1 of the low PIM core test module.

The detailed method is as follows: connect the power coupling ports P_C11 and P_C12 of the directional coupler to the input ports P_C1 and P_C2 of the power meter respectively, the test port TP1 of the low PIM core test module to the attenuator with known fixed attenuation amount and then to the calibration power meter. The signal source module generates two test signals f_1 and f_2. First, keep f_1 constant, and scan f_1 in the transmitting bandwidth with fixed frequency step (selected according to the demand), while measuring and recording the power of the signal source module output f_1, the measured power of the P_C1 port of the power meter and the measured power of the calibration power meter to form the first carrier channel attenuation data mapping table. Then, keep f_1 constant, scan f_2 with the same frequency step in the transmitting bandwidth, and simultaneously measure and record the output power of the signal source module f_2, the measured power of the P_C2 port of the power meter and the measured power of the calibration power meter to form the second carrier channel fading data mapping table. The measured attenuation data mapping table is stored in the IPC and the system control software module, and the real-time feedback control of the signal source module and the real-time display control of the power meter are realized by calling the data mapping table during the test. The calibration process is done automatically through software control.

Receiving PIM signal power calibration is to ensure that the spectrometer (or real-time detection and analysis module) measured PIM signal power for the actual PIM signal power generated at the device end to be tested, that is, the need to obtain the sum of link loss for reflection, transmission and extended port reception calibration between a low PIM core test module test ports TP1, TP2 and TP3 to the PIM product receiving module output port PIM_1 and PIM_2, respectively.

The detailed method is as follows: first, the reflection, transmission and extended port reception calibrations are performed for test ports TP1, TP2 and TP3 of the low PIM core test module to the output port PIM_1 of the PIM receiver module. Signal source module output signal f_pim is input into the low PIM core test module test port TP1, PIM product receiver module output port PIM_1 is connected to the calibration power meter, the signal source scan f_pim in the reception bandwidth with 1 MHz frequency step, while measuring the signal source output power and power measured by calibration power meter to form the reflection PIM receive channel attenuation data mapping table, so as to complete the reception calibrations of the reflection and expansion port. Similarly, the transmission and expansion port reception calibration can be completed for the signal source module output signal f_pim, successively into the low PIM core test module test ports TP2 and TP3, after the same process. Similarly, the reflection, transmission and extended port reception calibration of the test ports TP1, TP2 and TP3 of the low PIM core test module to the output port PIM_2 of the PIM receiver module can be completed. The measured attenuation data mapping table is stored in the IPC and the system control software module, and the data mapping table is called during the testing process to realize the real-time display control of the spectrometer or the real-time detection and analysis module. The calibration process is completed automatically through software control.

5.3.2 Broadband High Sensitivity Passive Intermodulation Measurement Based on Bridge and Filter Combination Method

5.3.2.1 Main Technical Content

The specific technical scheme of transmissive PIM measurement and reflective PIM measurement is shown in Figures 5.8 and 5.9, respectively, including: signal source 1, signal source 2, power amplifier 1, power amplifier 2, directional coupler 1, directional coupler 2, filter f_1, filter f_2, combiner, transmit filter, first 3 dB bridge, low-pass filter 1, low-pass filter 2, second 3 dB bridge, receiving filter 1, receiving filter 2, absorption load, low-noise amplifier, spectrometer, band-pass filter and high-power load.

Signal source 1 and signal source 2: to generate two test signals $f_1, f_2, f_1 < f_2$.

Power amplifier 1 and power amplifier 2: for power amplification of the test signal to obtain high-power carrier signals f_1, f_2.

Directional coupler 1 and directional coupler 2: to couple output two carrier signals and to detect carrier signal power.

Filter f_1: a high-power low-pass (or band-pass) filter, to filter the first carrier signal f_1 to remove harmonics and spurious waves generated by the previous link.

Filter f_2: a high-power low-pass (or band-pass) filter, to filter the second carrier signal f_2 to remove the harmonics and spurious waves generated by the previous link.

Combiner: to combine the two carrier signals f_1 and f_2 into one test carrier signal.

Transmit filter: a high-power low PIM bandpass filter with $(f_1+f_2)/2$ of the filter center frequency and ≤ -135 dBm@2×43 dBm for the third-order PIM value, which is usually required for a suppression of more than 80 dB at $2f_1-f_2$ and $2f_2-f_1$ out-of-band, to filter the synthesized carrier signal to remove other intermodulation and spurious components, and to ensure the purity of the carrier signal entering the DUT.

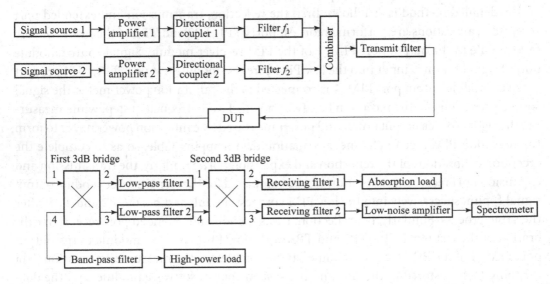

FIGURE 5.8   Transmission PIM measurement.

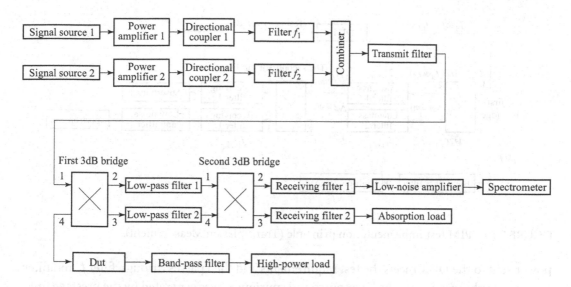

FIGURE 5.9 Reflective PIM measurement.

The first 3 dB bridge: a high-power low PIM 90° orthogonal bridge with ≤−135 dBm@2×43 dBm for the 3rd-order PIM value, to achieve the power distribution and synthesis of the carrier and PIM signal.

Low-pass filter 1 and low-pass filter 2: to reflect the carrier signal through the PIM signal. A high-power low PIM low-pass filter which can pass PIM signal $2f_1-f_2$, $3f_1-2f_2$, $4f_1-3f_2$... is required. Generally, it is required that the suppression at f_1 and f_2 in the stop band is more than 100 dB, and its own 3rd order PIM value is ≤−135 dBm@2×43 dBm. The two filters are exactly identical, and strict consistency must be ensured, and the phase shift difference is less than 0.5°.

The second 3 dB bridge: to realize the power synthesis of PIM signal.

Receiving Filter 1 and Receiving Filter 2: to further filter the PIM signal and suppress other spurious interference. Both are identical.

Absorbing load: to absorb small signals from leaks.

Low noise amplifier: to amplify the PIM signal.

Spectrometer: to detect PIM signals.

Band-pass filter: a high-power low PIM band-pass filter with $(f_1+f_2)/2$ of the filter center frequency, which can pass the carrier signal f_1 and f_2 and reflect the PIM signal, to further filter the carrier signal after the test. It is required that the suppression at $2f_1-f_2$ and $2f_2-f_1$ in the out-of-band is more than 80 dB, and its own 3rd order PIM value is ≤−135 dBm@2×43 dBm.

High power load: a low PIM high-power load, to absorb the carrier signal after the test. It is usually required that its own 3rd-order PIM value is ≤−135 dBm@2×43 dBm.

Signal source 1 and signal source 2 generate test signals f_1 and f_2, respectively, which is amplified through power amplifier 1 and power amplifier 2 to obtain high-power test carrier signals f_1 and f_2. The two carrier signals f_1 and f_2 are passed through directional coupler 1 and directional coupler 2, respectively, to achieve power coupling detection and ensure that the

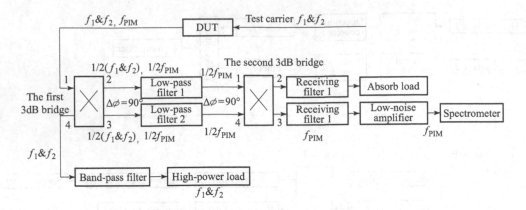

FIGURE 5.10   PIM test implementation principle (Transmission Measurement).

power sent to the DUT meets the test requirements and then passes through filter f_1 and filter f_2, respectively, to suppress the harmonics and spurious waves generated by the prestage link. After filtering, the two carrier signals f_1 and f_2 are synthesized through the combiner, and the synthesized carrier signal is passed through the transmit filter to filter out other frequency signals other than the test carrier signal to obtain the pure f_1 and f_2 test carrier signals.

As shown in Figures 5.8 and 5.10, when transmissive PIM measurement is performed, the test carrier signal output from the transmit filter is input to the DUT, and the DUT generates a mixed signal of carrier and PIM, which enters the first 3 dB bridge port 1 and outputs the power distribution from ports 2 and 3, where the signal phase difference between ports 2 and 3 is 90°, and the mixed signal passes through the low-pass filter 1 and low-pass filter 2, respectively. Hereafter, the test carrier signal is reflected back to the first 3 dB bridge, output from port 4 of the first 3 dB bridge, filtered by band-pass filter and absorbed by the high power load. The two PIM signals enter ports 1 and 4 of the second 3 dB bridge with 90° of the phase difference, which are synthesized and output from port 3 of the second 3 dB bridge, further suppressed by receiving filter 2, and then detected by the spectrometer after low noise amplifier; port 2 of the second 3 dB bridge is connected to the absorption load after passing through receiving filter 1.

The principle of reflective PIM detection is similar to that of transmissive PIM detection but the connection position of the DUT. As shown in Figure 5.9, the test carrier signal output from the transmit filter is input to port 1 of the first 3 dB bridge, and the DUT is connected between port 4 of the first 3 dB bridge and the bandpass filter, and the carrier signal is reflected by low-pass filter 1 and low-pass filter 2 at ports 2 and 3 of the first 3 dB bridge after passing through the first 3 dB bridge and output from port 4 of the first 3 dB bridge to the DUT. The reflected PIM signal generated by the device enters the first 3 dB bridge from port 4, and the power distribution is output from ports 2 and 3 of the first 3 dB bridge; at this time, the phase difference of the output signals of ports 2 and 3 is −90°, and they enter the ports 1 and 4 of the second 3 dB bridge after passing through low-pass filter 1 and low-pass filter 2 respectively, and the phase difference of the two PIM signals at ports 1 and 4 is −90°, and after synthesizing from Port 2 of the second 3 dB bridge output to further suppress the spurious through the receiving filter 1, and then passing through the low noise amplifier to detect by the spectrometer, the second 3 dB bridge port 3 is connected to the absorb load through the receiving filter. The test carrier signal is filtered through the bandpass filter and then absorbed by the high power load.

When switching between the transmissive PIM measurement method and the reflective PIM measurement method, in addition to changing the connection position of the DUT, the positions of the absorption load and the low-noise amplifier and spectrometer need to be swapped.

In this section, a method is proposed to achieve broadband PIM measurement using a 90° phase 3 dB bridge combined with a low-pass filter, which cleverly uses the phase and signal synthesis characteristics of the 90° phase 3 dB bridge, in combination with a low-pass filter, effectively solves the contradiction between high isolation and wide bandwidth in traditional PIM detection, using the broadband performance of the bridge to ensure the passage of broadband signals, while the carrier and PIM signal separation is achieved by using low PIM low-pass filter; through such a combination, the low-pass filter out-of-band suppression system can be improved to achieve high isolation between the transceiver (PIM) and transmitter (carrier) channels, so as to avoid the problem that it is difficult to achieve the technical specifications of the duplexer when the traditional method is used to achieve broadband measurement, effectively reducing the difficulty of design. Moreover, all the key components in the scheme can be realized through the waveguide structure, which can make the system as a whole to obtain more stable low residual PIM performance and achieve broadband high sensitivity PIM test.

Based on this scheme, we have designed and implemented an S-band broadband high-sensitivity PIM test system with a waveguide structure, which can simultaneously achieve 3rd to 15th order PIM tests when the carrier is 2160 and 2210 MHz. The system itself has a 3rd order residual PIM < -130 dBm@2×43 dBm and a 7th order residual PIM < -150 dBm@2×43 dBm, which simultaneously has a wide receiving band, high power and high sensitivity performance, which verifies the effectiveness of the scheme.

5.4 JOINT TIME DOMAIN-FREQUENCY DOMAIN REAL-TIME DETECTION ANALYSIS OF PASSIVE INTERMODULATION

When conducting PIM measurements under general requirements, commercial spectrometers can indeed meet the requirements, but there are some problems when conducting high sensitivity and high dynamic range tests. The main reasons are as follows.

1. The PIM signal is particularly weak, usually in the order of -150 dBm.

2. The moment of PIM signal appearance is uncertain, and the period of conducting a test is relatively long, which requires long-time observation records.

3. PIM signals are usually transient burst, may be short in duration, and are of variable duration.

4. Single target PIM signal bandwidth is relatively narrow, usually a single frequency point observation, requiring a spectral resolution of Hz level.

When using a commercial spectrometer for testing, the test sampling period will be extended as the scan time of the spectrometer increases. In actual components and systems, there are many transient jump PIM signals, especially in the temperature cycling conditions; such transient jump PIM signals are short in duration but large in amplitude, which are sufficient to cause interference to communications, and the traditional spectrometer frequency domain measurement means for such sudden PIM signals cannot be fully detected. Because the transient jump PIM signal in the time domain is a pulse signal, while in the frequency domain presents a spectrum spreading, it cannot be in the frequency domain to achieve accurate testing. Therefore, commercial spectrometers cannot meet the needs of high-performance PIM comprehensive detection. Moreover, the PIM signal itself is very weak, and mixed with noise, the general oscilloscope cannot directly detect it in the time domain.

In order to study the characteristics of PIM in depth, it is necessary to analyze and process the actual PIM data. The measurement of PIM has the characteristics of time-consuming and labor-intensive. Therefore, it is necessary to complete the full-bandwidth recording of real-time data and the postprocessing function of the data while conducting real-time data acquisition.

For the transient PIM characteristics such as low level, suddenness and short duration, this section proposes a PIM measurement method based on joint real-time detection in the time domain-frequency domain, real-time acquisition of the original PIM signal and real-time processing in the time domain-frequency domain, to effectively capture, detect and analyze transient jump PIM signals in the time domain while achieving high sensitivity PIM measurement in the frequency domain. Compared with the traditional spectrometer measurement method, this measurement method has higher detection sensitivity and better detection real-time, can achieve more accurate measurement and meet the postprocessing requirements.

5.4.1 Principle and Procedure of Joint Time-frequency Domain Passive Intermodulation Real-Time Detection and Analysis System

The principle of the joint time-frequency-domain PIM real-time detection and analysis system is shown in Figure 5.11, which mainly includes RF processing, IF processing, digital signal processing and software processing modules.

The PIM RF signal acquisition processing specifically includes the following processes:

1. Low noise amplification 1: low noise amplification of the input PIM small signal with a very low noise factor to ensure high sensitivity reception.

2. ESC filtering: the filter center frequency is set by an adjustable filter according to the frequency of the PIM signal to be detected, so as to filter the PIM signal.

3. First-stage down conversion: the PIM signal is down-converted to the first IF 321.4MHz.

4. First IF filtering: band-pass filtering of the first IF after down-conversion is performed to filter out other spurious components.

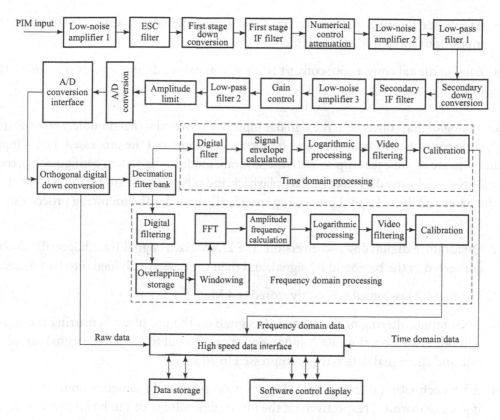

FIGURE 5.11 Principle of joint time-frequency-domain PIM real-time detection and analysis system.

5. Digital attenuation: together with the poststage gain control, the automatic gain control function is implemented to adjust according to the dynamic range of the input PIM signal.

6. First-stage IF amplification (low noise amplification 2): amplify the first IF signal.

7. Low-pass filtering 1: filter the first IF after amplification to remove its harmonic interference.

8. Secondary down conversion: down-convert the first IF to the second IF at 21.4 MHz.

9. Second IF filtering: band-pass filtering of the second IF after down-conversion to filter out other spurious components.

10. Second IF amplification (low noise amplification 3): amplify the second IF signal.

11. Gain control: together with the prestage NC attenuation, the automatic gain control function is implemented to adjust according to the dynamic range of the PIM signal.

12. Low-pass filtering 2: filter the second IF after amplification to remove its harmonic interference.

13. Amplitude limiting: protect the back-end analog/digital converter circuit when the input signal amplitude is too large.

14. Analog/digital conversion: convert IF analog signals to digital signals at a rate of 128 MS/s.

After analog/digital conversion, the signal is input to the digital signal module, and real-time analysis is performed in both time and frequency domains, and the processed data is input to the software module for output. With the software module, the system configuration, real-time display, real-time data storage, data playback and other processing are implemented.

The baseband digital signal processing specifically includes the following processes:

1. Quadrature digital down-conversion: The 21.4 MHz IF signal is orthogonally down-converted to the baseband IQ signal, and then the data are divided into two paths.

2. The first IQ raw signal is directly stored at 4 MS/s.

3. Decimation filtering of the second IQ signal: on the premise of ensuring the signal without distortion, the data compression is carried out to reduce the signal sampling rate and the signal data rate is compressed to 80 kS/s.

4. After decimation filtering, time domain processing and frequency domain processing are performed, respectively: if the minimum pulse width to be analyzed is T, the analysis bandwidth $fs \geq 4/T$.

The detailed processing in the time domain or frequency domain is as follows:

1. Time domain process

   1. Digital filtering: set the filtering bandwidth, $B = 1/Tmin$, according to the minimum width of the time domain of the PIM pulse to be detected; $Tmin$ is the minimum duration of the PIM time domain pulse to be detected.

   2. Signal time domain envelope calculation and logarithmic processing: $20lg|I+jQ|$, to obtain the time domain envelope amplitude.

   3. Video filtering: improve waveform display.

   4. Calibration: complete data calibration and alignment to get the final time domain processed results.

2. Frequency domain process

   1. Digital filtering: frequency domain digital filtering according to the analysis bandwidth.

   2. Overlap storage: overlap storage of IQ signal at 87.5% overlap rate to ensure no distortion after data windowing.

3. Windowing processing: the IQ signal is processed by windowing. The window functions include Kaiser, Rectangular, Hamming and Hanning. The algorithm of windowing is a dot product matrix that multiplies the IQ signal with a prestored window function.

4. Fast Fourier transform (FFT): perform a 32,768-point fast Fourier transform.

5. Magnitude and frequency calculation and logarithmic processing: $20 \lg |IFFT + jQFFT|$, to obtain the frequency domain amplitude.

6. Video filtering: improve waveform display.

7. Calibration: complete data calibration and alignment to obtain the final frequency domain processed results.

The processed time-domain data and frequency-domain data are output as the comprehensive time-frequency-domain analysis results of PIM by the control software through a high-speed data interface.

The relation between the detectable minimum PIM signal amplitude and the shortest duration T_{min} in the time domain is

$$S_{min\_t} = -174 + NF + 10 \lg B + SNR_p = -174 + NF + 10 \lg(1/T_{min}) + SNR_p$$

where S_{min_t} is the time domain detection sensitivity; SNR_p is the signal-to-noise ratio required by the effective detection of PIM signals, usually 10 dB; NF is the link noise factor.

The frequency domain detection sensitivity S_{min_f} is

$$S_{min\_f} = -174 + NF + 10 \lg RBW + SNR_p$$

where RBW is the system resolution bandwidth.

From the above process, it can be seen that the joint time-frequency domain detection analysis of PIM can not only achieve high sensitivity and real-time detection but also effectively detect the transient jump PIM signal.

5.4.2 Key Technical Specifications of Joint Passive Intermodulation Real-time Detection and Analysis System in Time Domain and Frequency Domain

1. Real-time analysis in time and frequency domain

The illustration block diagram of real-time processing is shown in Figure 5.12, where M_i ($i = 0, 1, 2,...$) is the data of each frame stored in the memory, $CALC_i$ ($i = 0, 1, 2,...$) is the time to process the data of each frame, including calculation processing time, display refresh time, etc. Only when $CALC_i$ is less than the time to process the data of each frame, it is real-time processing, otherwise it is non-real-time processing.

Before the FFT operation, the windowing process is required. In order to avoid data loss due to the windowing process and to maintain the amplitude accuracy, overlapping

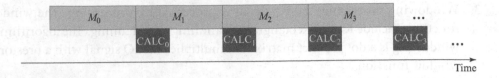

FIGURE 5.12   Block diagram of real-time processing.

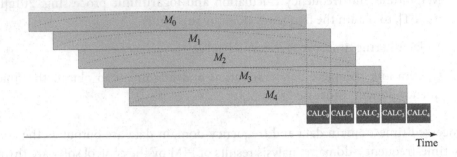

FIGURE 5.13   Overlapping storage.

storage is required first. 90% overlapping is used in this method design, when $CALC_i$ should be less than 10% of the time per frame data, as shown in Figure 5.13.

In the design, the processing of the signal is carried out in the time and frequency domains in parallel, so that the time and frequency domain information of the signal can be obtained simultaneously.

2. Frequency range

Based on the actual detection demand, the system working frequency range index requirement developed in this chapter is 1~18 GHz. The system RF unit adopts the ultradyne secondary frequency conversion scheme; after the first-level mixing, the signal is changed into the first-level IF signal of 321.4 MHz, and after the second level mixing, the first level IF signal of 321.4 MHz is changed into the second level IF signal of 21.4 MHz. The first-level mixing uses a high fundamental oscillation signal, so, as long as the frequency range of the first-level fundamental oscillation signal is 1.3214~18.3214 GHz, it can meet the system working frequency range index requirements.

3. Maximum analysis bandwidth

Based on the actual detection requirements, the maximum analysis bandwidth of the system requires at 32 kHz, and the data rate of 32 kS/s is selected for the FFT operation processing, which can meet the requirements of the maximum analysis bandwidth of the system.

4. Minimum resolution bandwidth

Based on the actual detection requirements, the system minimum resolution bandwidth index requirement is 1 Hz. the system resolution bandwidth RBW is calculated by the formula

TABLE 5.1   Correspondence of Window Function and Window Factor

Window Function	Window Factor
Kaiser	2.23
Rectangular	0.89
Hamming	1.30
Hanning	1.44

$$\text{RBW} = \frac{W_f \times f_s}{N}$$

where W_f is window factor, f_s is sampling frequency, and N is number of points of FFT.

The window factor W_f is related to the type of window function used for windowing to the data, and the correspondence between common window functions and window factors is shown in Table 5.1.

In Table 5.1, the minimum and maximum value of the window factor W_f are 0.89 and 2.23, respectively.

Choosing $f_s = 32\,\text{kHz}$ and $N = 32768$, the minimum value RBW_{min} and the maximum value RBW_{max} of RBW can be obtained with the value of window factor W_f

$$\text{RBW}_{min} = (0.89 \times 32000) / 32768 = 0.87 (\text{Hz})$$

$$\text{RBW}_{max} = (2.23 \times 32000) / 32768 = 2.18 (\text{Hz})$$

From the above analysis results, it can be concluded that the system has the smallest resolution bandwidth of 0.87 Hz when the window function type is rectangular, $f_s = 32\,\text{kHz}$, and $N = 32{,}768$, and the largest resolution bandwidth of 2.18 Hz when the window function type is Kaiser, $f_s = 32\,\text{kHz}$, and $N = 32768$.

In summary, the minimum resolution bandwidth of the system of 1 Hz can be met by choosing the window function type rectangular, $fs = 32\,\text{kHz}$ and $N = 32768$.

5. Minimum analyzable time domain pulse width

Based on the actual detection requirements, the minimum analyzable time domain pulse width of the system is 125 µs. From the time domain perspective, four times sampling of the signal can completely characterize the signal, so that the time domain data sampling period $T_s = 31.25$ µs, i.e., the sampling frequency $f_s = 32\,\text{kHz}$, and the requirement of the minimum analyzable time-domain pulse width is 125 µs. From the frequency domain perspective, when the time domain pulse width is 125 µs, its 3 dB bandwidth is 8 kHz. In the design, the time domain signal processing signal bandwidth is limited to 8 kHz to meet the requirements. Since the signal is a pulse signal, it occupies a certain signal bandwidth in the frequency domain. When the pulse width in the time domain is T (s), its 3 dB bandwidth B (Hz) is about

$$B = 1/T$$

TABLE 5.2   Correspondence between Bandwidth S and P_{min}, PW

P_{min}/dBm	B/Hz	PW/ms
−120	4000	0.25
−130	400	2.5
−140	40	25
−155	1.26	794

Therefore, the minimum power of the pulse signal P_{min} that the system can analyze is

$$P_{min} = -174\,\text{dBm} + 10\lg\frac{B}{1\,\text{Hz}} + \text{NF} + N_{slack},$$

where B is signal bandwidth, its value is related to the time domain pulse width; NF is link noise factor, selecting 8 dB; N_{slack} is margin, selecting 10 dB.

The result is as follows:

$$P_{min} = -174\,\text{dBm} + 10\lg\frac{B}{1\text{Hz}} + 8\text{dB} + 10\text{dB} = -156 + 10\lg B$$

The P_{min}, time domain pulse width PW corresponding to different bandwidths B are shown in Table 5.2.

6. Working dynamic scope

   Based on the actual testing requirements, the dynamic range of the system is required to be greater than or equal to 70 dB. The dynamic range of the system is generally composed of two parts: one is the attenuation range of the controllable attenuator of the analog link; the other is the dynamic range of the ADC (analog-to-digital converter) under the condition of meeting the system test accuracy.

   In the design, the attenuation range of the controllable attenuator in the analog link is 40 dB, and the dynamic range of ADC is 45 dB, which can meet the requirement that the dynamic range of the system is greater than or equal to 70 dB.

7. Minimum input signal power

   The power requirement of the minimum input signal is less than or equal to −155 dBm. The calculation formula of the display average noise level (DANL) of the system is

$$\text{DANL} = -174\text{dBm} + 10\lg\frac{\text{RBW}}{1\text{Hz}} + \text{NF}$$

   When $\text{RBW} = \text{RBW}_{max} = 2.18\,\text{Hz}$ and $\text{NF} = 8\,\text{dB}$, the maximum value of the system display average noise level DANL_{max} can be obtained as

$$\text{DANL}_{max} = -174\text{dBm} + 3.4\text{dB} + 8\text{dB} = -162.6\text{dBm}$$

FIGURE 5.14 Time-frequency domain joint PIM real-time detection and analysis system experimental prototype.

When RBW = RBW$_{min}$ = 0.87 Hz and NF = 8 dB, the minimum value of the system display average noise level DANL$_{min}$ can be obtained as

$$DANL_{min} = -174dBm - 0.6dB + 8dB = -166.6dBm$$

In conclusion, the maximum average noise level of the system is −162.6 dBm, so the minimum signal power that the system can accurately test is −155 dBm.

5.4.3 Prototype Development and Experimental Test Result Analysis

Based on the principle approach described in the previous section, an experimental prototype is developed in this chapter as shown in Figure 5.14, with both time domain and frequency domain measurement window options.

For specific detection needs, the minimum time-domain pulse to be detected is 125 μs, with the analysis bandwidth f_s = 32 kHz, the window function rectangular, the window factor W_f = 0.89, the corresponding frequency-domain resolution bandwidth RBW = $W_f \times$/ f_s/32 768 = 0.87 Hz and the time-domain filtering bandwidth B = 1/125 μs = 8 kHz. When the link noise figure NF = 8.5 dB and SNR$_p$ = 10 dB, the time domain sensitivity Smin_t and the frequency sensitivity Smin_f can be obtained

$$S_{min\_t} = -174 + NF + 10\lg B + SNR_p = -116.5dBm$$
$$S_{min\_f} = -174 + NF + 10\lg RBW + SNR_p = -155.5dBm$$

That is, when the amplitude is greater than −116.5 dBm, the jump PIM with the minimum duration of 125 μs can be effectively detected.

In order to verify the effectiveness of the proposed method, the PIM of a 7/16 DIN type connector is tested on a high-sensitivity PIM research platform under temperature cycling

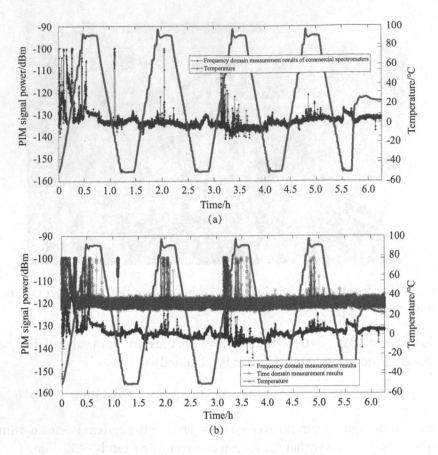

FIGURE 5.15   PIM test results. (a) Commercial spectrometer detection results. (b) Time-frequency domain joint PIM real-time detection results.

conditions, in which five temperature cycling cycles are set within a temperature change range of −50°C~90°C and at a temperature change rate of 5°C/min, which is maintained at high and low temperatures for 30 min. By reflective measurement, the output 3rd order PIM signal is divided into two equal ways through the power divider and then connected to the commercial spectrometer and the time domain frequency domain joint PIM real-time detection and analysis system, respectively, to measure synchronously. After 6h measurement, the obtained results are shown in Figure 5.15.

As can be seen from the test results shown in Figure 5.15, the trends of the test results of these two methods are basically the same, but it is obvious that the detection results of the joint time- frequency-domain PIM real-time detection method are more accurate. In order to distinguish the difference between these two methods more clearly, so a small section of the entire test data was intercepted for local observation after zooming in, as shown in Figure 5.16. It can be seen that the frequency domain detection results of these two methods overlap completely, thus verifying the accuracy of the new method. In addition, the time-domain detection captures the transient jump PIM signal, which cannot be detected by the general-purpose spectrometer. The joint time-frequency-domain PIM real-time detection method obtains a comprehensive result by processing and analyzing

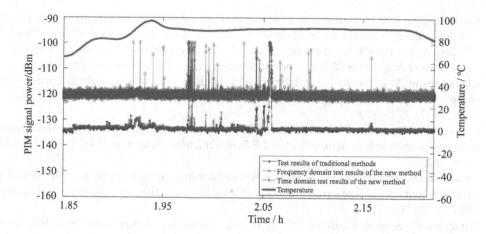

FIGURE 5.16   Partial comparison of the test results of time domain frequency domain joint PIM real-time detection method and the traditional spectrometer detection method.

the time-domain-frequency domain simultaneously, which has the advantages of both high sensitivity of frequency domain detection and real-time time domain detection. In addition, on the basis of the completion of high-sensitivity detection, it also ensures the effective capture of transient jump PIM signal, making up for the defects of the traditional method of leakage detection.

Compared with the detection using a traditional commercial spectrometer, the detection method provided in this section can obtain both frequency domain and time domain detection data, and the final detection results are obtained after comprehensive analysis. It can be clearly seen from the experimental results that the detection method can achieve more accurate testing, and on the basis of satisfying the frequency domain detection sensitivity, it can capture transient pulse signals that cannot be captured by traditional commercial spectrometers, with higher sensitivity and real-time performance, solving the defects of leakage detection of transient burst PIM pulse signals in the traditional detection method, significantly improving the accuracy and efficiency of PIM detection, effectively improving the detection sensitivity and real-time and providing a new way of thinking and approach for PIM detection.

BIBLIOGRAPHY

1. Ye M., He Y., Sun Q., et al. Review of passive intermodulation interference problem under high power signals. *Space Electronic Technology*, 2013, 10(1): 75–83.
2. Zhu H. *Practical RF Test & Measurement*. Beijing: Publishing House of Electronics Industry, 2012.
3. Kim J. T., Cho I. K., Jeonc M. Y., et al. Effects of external PIM sources on antenna PIM measurements. *Etri Journal*, 2002, 24 (6): 435–442.
4. Chen X., Cui W., Li J., et al. Measurement and localization of passive intermodulation distortion of high power microwave component for space application. *Space Electronic Technology*, 2015, 12 (6): 1–7.
5. Wilkerson J. R. Passive intermodulation distortion in radio frequency communication systems. Raleigh: North Carolina State University, 2010.

6. Wetherington J. M. Robust analog canceller for high - dynamic - range radio frequency measurement. *IEEE Transactions on Microwave Theory and Techniques*, 2012, 60 (6): 1709–1719.

7. Rawlins A. D., Petit J. S., Mitchell S. D. PIM characterization of the ESTEC compact test range. *Proceedings of the 28th European Microwave Conference*, Amsterdam, 1998, 544–548.

8. Christianson A. J., Henfie J. J., Chappell W. J. Higher order intermodulation product measurement of passive components. *IEEE Transactions on Microwave Theory and Techniques*, 2008, 56 (7): 1729–1736.

9. Golikov V., Hienonen S., Vainikainen P. Passive intermodulation distortion measurements in mobile communication antennas. *IEEE 54th Vehicular Technology Conference*, 2001, (1–4) 2623–2625.

10. Smacchia D., Soto P., Boria V. E., et al. Advanced compact setups for passive intermodulation measurements of satellite hardware. *IEEE Transactions on Microwave Theory and Techniques*, 2018, 66 (2): 700–710.

11. Smacchia D., Soto P., Guglielmi M., et al. Implementation of waveguide terminations with low - passive intermodulation for conducted test beds in backward configuration. *IEEE Microwave And Wireless Components Letters*, 2019, 29 (10): 1–3.

12. Shitvov A. P., Zelenchuk D. D., Schuchinsky A., et al. Mapping of passive intermodulation products on microstrip lines microwave symposium digest. *2008 IEEE MTT - s International.* IEEE, 2008: 1573–1576.

13. Gao F., Zhao X., Ye M., et al. A passive intermodulation measuring method based on the coupling of dipole near-field. *Space Electronic Technology*, 2018, 15 (3): 15–21.

14. Boyhan J. W., Lenzing H. F. Satellite passive intermodulation: systems considerations. *IEEE Transactions on Aerospace & Electronic Systems*, 2002, 32 (3): 1058–1064.

15. Shantnu M., Yves P., Pierre C. A review of recent advances in passive intermodulation and multipaction measurement techniques. *International Symposium on Antenna Technology & Applied Electromagnetics & Canadian Radio Sciences Conference.* IEEE, 2017: 1–4.

16. Carceller C., Soto P., Boria V. E., et al. Design of compact wideband manifold - coupled dnmultiplexers. *IEEE Transactions on Microwave Theory and Techniques*, 2015, 63 (10): 3398–3407.

17. Carceller C., Soto P., Boria V. E., et al. Design of hybrid folded rectangular waveguide filters with transmission zeros below the passband. *IEEE Transactions on Microwave Theory and Techniques*, 2016, 64 (2): 475–485.

CHAPTER 6

PIM Suppression Technology for Microwave Components

Tiancun Hu, He Bai, Qi Wang, and Lu Tian

CONTENTS

DOI: 10.1201/9781003269953-6

6.1 OVERVIEW

In the design, production and commissioning phases of communication systems, PIM interference control indexes need to be assigned. However, there is a lack of research related to the impact of PIM on the receiver performance of communication systems, so that the impact of PIM interference on the communication systems must be quantitatively analyzed. In this chapter, by studying the elements of PIM impact on communication systems, the relational framework between communication performance and PIM characteristics is constructed for transceiver duplex systems, and a quantitative analysis model of PIM impact on communication system performance is established, which provides both theoretical support and guidance for the PIM design of communication systems and key technical support for the design and performance evaluation of communication systems.

Contact nonlinearity is one of the causes of PIM. The metal connection of microwave components is an important source of PIM, and low-quality metal connection is an important factor causing PIM. However, the type of surface coating, coating thickness, welding joints, etc. are key factors affecting the quality of the metal connection, so the study of coating materials and thickness, etc. on the effect of PIM law can provide a theoretical basis for low PIM surface treatment technology for microwave components. The generation of PIM of high-power microwave components is directly related to the electric field strength of their internal PIM-prone parts. In this chapter, the design scheme and tuning structure of microwave components are optimized using a comprehensive optimization design method for high-power microwave components such as waveguide duplexers, coaxial filters and antenna feeders, and a design scheme for low-PIM high-power microwave components is proposed.

Due to the limitations of platform capacity and cost, the transceiver-shared antenna becomes the best choice, so the antenna PIM will become the technical bottleneck of the transceiver-shared antenna system, and the PIM index is very challenging. This chapter will introduce the PIM design rules for spacecraft antennas and give some typical examples of low PIM satellite-borne feeds at present. PIM suppression techniques for terrestrial mobile communication antennas are affected by cost factors and are not applicable in some design rules for spacecraft antenna PIM suppression, so this chapter focuses on low PIM welding techniques. At the end of this chapter, the system level suppression method of spacecraft payload PIM is briefly introduced, and it is emphasized that the more stringent PIM indicators in the future must be effectively suppressed through system-level methods.

6.2 PERFORMANCE IMPACT OF PIM ON COMMUNICATION SYSTEMS

In the transceiver communication system, the transmitted signal and the received signal pass through the transceiver duplexer, high-power cable and antenna feed source at the same time. If the PIM products generated between different carriers of the transmitted signal fall into the receiving band, the PIM products will enter the receiver together with the received signals at the same time, forming an interference signal. This problem cannot be solved by using traditional filter and isolation methods. When the PIM level is low, the bottom noise of the received signal will be raised, so that the receiver signal-to-noise ratio decreases and the bit error rate increases; when the PIM level increases further, it will affect the normal operation of the entire communication system, and in serious cases, the PIM will flood the received signal, resulting in channel blockage and communication interruption, which will paralyze the entire communication system.

6.2.1 Influence Factors of PIM on Communication Systems

When PIM interference is superimposed on the uplink received signal, the Error Vector Magnitude (EVM) of the uplink received signal will be deteriorated so that the verdict of the uplink received signal will be affected and the error code elements will appear in the uplink signal transmission and then affect the demodulation performance of the uplink communication link. For the impact of PIM interference on the demodulation link of the communication system, the BER of the uplink transmission is used to measure.

The influencing factors of PIM interference on the communication system include the degree of device nonlinearity, downlink signal characteristics, and uplink signal characteristics. The correlation between the factors of PIM interference on the communication system and the performance of the communication system are shown in Figure 6.1.

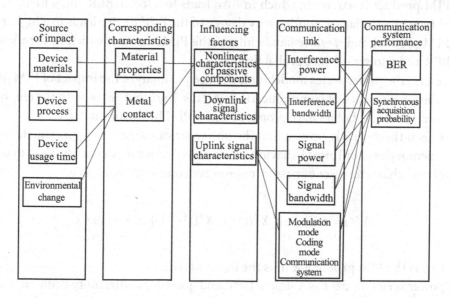

FIGURE 6.1 Correlation between the influencing factors of PIM interference on the communication system and the performance of the communication system.

The device material, process, usage time and environmental changes of passive microwave components can all affect their linear characteristics, that is, it is easy to lead to nonlinear effects. Passive nonlinearity can be mainly divided into material nonlinearity and contact nonlinearity. The nonlinear characteristics of passive devices, downlink signal characteristics, and uplink signal characteristics will influence the interference power, interference bandwidth, signal power and signal bandwidth in the communication link, which in turn affects the SIR (i.e., the ratio of useful signal to PIM interference signal power) in the communication link. The variation of the SIR in the communication link and the modulation mode of the uplink signal and other characteristics will affect the communication system performance indexes such as the BER and the synchronous capture ratio of the communication system.

This section will analyze the impact of PIM interference on the performance of communication systems, starting from the nonlinear characteristics of passive devices, downlink signal characteristics and uplink signal characteristics.

6.2.1.1 Correlation between the Nonlinear Characteristics of Passive Devices and the Performance of Communication Systems

There are two main causes of PIM interference in microwave components, namely contact nonlinearity and material nonlinearity. Contact nonlinearity is the nonlinearity caused by contacting with nonlinear current (or nonlinear voltage) behavior; material nonlinearity is the nonlinearity caused by the nonlinear properties of the material itself.

Since the PIM interference on the satellite is mainly caused by the nonlinearity of high-power downlink signals passing through passive devices, the nonlinearity degree of passive devices has a very large impact on the PIM products. When the nonlinearity degree of the PIM device is relatively small, the power of the PIM product is smaller, and the bandwidth of the PIM product is narrower, which in turn leads to a lower BER and a higher capture probability; while the larger nonlinearity degree of the PIM device leads to a higher power of the PIM product and a wider bandwidth of the PIM product, which in turn leads to a larger BER and a lower capture probability.

At present, the power series model is mostly used to analyze the impact of PIM interference on the performance of communication systems. In recent years, with the emphasis on PIM research, some scholars have proposed a new PIM behavior model.

It has been theoretically proven that the power series model can approximate any continuous memoryless nonlinear behavior, so the power series model can be used to describe the nonlinear characteristics of passive microwave components, namely

$$Y(t) = a_1 X(t) + a_2 X^2(t) + a_3 X^3(t) + \cdots + a_n X^n(t) + \cdots \tag{6.1}$$

where $Y(t)$ is the PIM product; $X(t)$ is the input signal.

The power series model has a simple form and good generalizability compared to other nonlinear models. Therefore, in this section, the power series model is used as the main model for analytical study.

Jacques Sombrin et al. in France proposed a new multicarrier PIM interference representation in France in 2014. This model is summarized from experimental results and is simpler than the power series model, which can effectively represent PIM interference and facilitate theoretical analysis. It states that the nonlinear behavior of PIM interference can be expressed as

$$v_{out} = \alpha \cdot v_{in} \cdot \left| v_{in} \right|^{\beta} \tag{6.2}$$

where v_{out} is the output signal (the output signal after the passive device contains the PIM), α is the linear parameter, v_{in} is the input signal, and β is the power parameter, whose value is between 0.6 and 1.5.

6.2.1.2 Correlation between Downlink Signal Characteristics and Communication System Performance

Because the frequency of PIM product is a linear combination of input signal frequency, the power of downlink signal, the bandwidth of downlink signal and the number of multiplexing channels of downlink signal will affect the power and bandwidth of PIM interference.

a. Correlation between downlink signal power and communication system performance

The power of the PIM interference signal is closely related to the power of the downlink signal. When the power of the downlink signal is relatively small, the power of the PIM interference signal is relatively small and can often be ignored. But when the downlink signal is transmitting at high power, the power of the PIM interference signal often cannot be ignored. The higher the power of the downlink signal, the higher the power of the PIM interference signal, and the power of the PIM interference signal increases nonlinearly with the increase of the downlink signal power. Taking the 5th power term in the typical power series model x^5 as an example, when the power of downlink signal increases by 1 dB, the power of PIM product generated by this term increases by 5 dB. It can be seen that the power of interference signal is closely related to the power of downlink signal.

In the case of a certain useful signal power, with the increase of the power of the PIM interference signal, the SIR of the entire communication link, the demodulation performance and synchronous capture performance of the communication system and the capture probability will be decreased, and the BER will be increased.

b. Correlation between downlink signal bandwidth and communication system performance

The spectral bandwidth of the downlink signal has an important impact on the interference bandwidth, and the bandwidth of the PIM product increases with the increase of the PIM order. The spectrum of PIM products in the dual-carrier case is shown schematically in Figure 6.2.

The wider the spectral bandwidth of the downlink signal, the wider the bandwidth of the interfering signal. Due to the spectral characteristic that the higher the PIM order, the wider the bandwidth, the wider spectrum of the downlink signal will cause more orders to fall into the receiving band of interest. Table 6.1 illustrates the order of

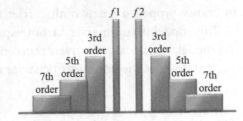

FIGURE 6.2   Frequency spectrum of PIM products in the case of dual carriers.

TABLE 6.1   Statistics of the Order of the PIM Products Generated by the Downlink Transmit Signal with Different Bandwidths Falling into the Receiving Band

Downlink Center Frequency1/GHz	Downlink Center Frequency2/GHz	Uplink Center Frequency/GHz	Uplink Receiving Bandwidth/MHz	Downlink Bandwidth/MHz	Order of PIM Falling into the Receiving Band
2.17	2.20	2.049	3	2	9
				5	9, 11
				10	7, 9, 11, 13
				12	7, 9, 11, 13, 15

PIM products generated by downlink transmitted signals with different bandwidths falling into the uplink receiving band in the dual-carrier case.

As can be seen from the statistics in Table 6.1, in the case of the same downlink frequency, uplink frequency and uplink receiving bandwidth of the signal, the wider the bandwidth of the downlink signal, there will be more PIM products falling into the receiving band of the uplink signal. This indicates that the bandwidth of the downlink signal directly affects the interference power and interference bandwidth of PIM interference, which affects the SIR of the uplink communication link and then affects the BER and correct capture probability of the uplink of the communication system.

c. Correlation between the number of downlink signal multiplexing paths and communication system performance

The number of multiplexed downlink signals has an impact on the bandwidth of PIM interference signals and the power of PIM interference signals. Some literature studies show that the spectrum complexity of PIM products increases exponentially with the increase of carrier number and product order. Figure 3.1 shows the simplified traversal tree of the product frequency of the 5th order PIM under three carriers.

Although the frequency traversal tree of the 5th order PIM product under three carriers given in Figure 3.1 has been simplified by the method of number theory, it is still relatively complex. As the number of carriers and the order of PIM products increase, the traversal tree of PIM products will be more complex. The frequency components of PIM interference will be more, and the spectrum will be more complex. This indicates that the number of multiplexed paths of the downlink signal affects the interference power and interference bandwidth of PIM interference, which affects the SIR of the uplink communication link, the BER and capture probability of the uplink of the communication system.

6.2.1.3 Correlation between Uplink Signal Characteristics and Communication System Performance

The uplink signal characteristics that have an impact on the performance of the communication system include signal power, signal bandwidth, modulation mode, coding method and communication system (fixed frequency spread spectrum).

The power of the uplink signal directly affects the SIR of the uplink communication link. When the interference is constant, the higher the uplink signal power, the higher the SIR. The higher the SIR, the better the performance of the communication system, and the lower the BER, the higher the capture probability.

The bandwidth of uplink signal also affects the performance of the communication system. PIM interference generally refers to the PIM products falling into the uplink receiving band, when the uplink receiving power is constant and the receiving band becomes wider, the power of the PIM interference falling into the receiving band becomes higher, the SIR will be decreased, the performance of the communication system becomes worse, the BER becomes higher, and the probability of correct synchronization capture will be decreased.

The modulation mode, coding method and communication system of uplink signal will have an impact on the performance of the communication system. The signal will be modulated by different modulation modes to get different constellation diagrams, which are directly related to the antiinterference performance of the communication system, so the impact of PIM interference on the performance of the communication system is different under different modulation modes, and the BER and correct synchronization capture probability of the uplink received signal is also different. The uplink signal is coded, and the error correction capability of the communication system is different for different coding methods. Therefore, the impact of PIM interference on the communication system is different for different coding methods, and the coding method affects the BER of the uplink received signal. Spread spectrum communication has the characteristics of antimultipath and antiinterference compared with fixed frequency communication, whether the spread spectrum communication system is used will affect the performance of the communication system under PIM interference, i.e., affect the BER and capture probability of the communication system.

The uplink signal power, bandwidth, modulation mode, coding method and communication system are the factors that influence the performance of PIM on the communication system. Under the appropriate modulation, coding and communication system, the higher the SIR of the uplink, the better the performance of the communication system, and the lower the BER of uplink signal transmission the higher the capture probability.

6.2.2 Analysis of the Impact of PIM on the Demodulation Link of Communication Systems

In this subsection, the impact of PIM interference on the demodulation performance of the communication system will be analyzed according to the statistical signal processing theory in combination with the characteristics of PIM interference. In the following analysis model of the effect of PIM interference on the demodulation performance of communication system, the analysis of the effect of PIM interference on the demodulation link under constant SIR and the analysis of the effect of PIM interference on the demodulation link under time-varying SIR are separately described according to whether the SIR is constant or not.

6.2.2.1 Analysis of the Effect of PIM Interference on Demodulation Process under Constant SIR

The signal characteristics of PIM interference at the uplink receiving frequency and then the variation curve of the BER with the SIR are analyzed under a certain constant SIR, combined with the power series model and the downlink transmit signal characteristics. The effect of PIM interference on the demodulation link at constant SIR is analyzed in a Gaussian channel, in which the power of the uplink useful received signal is constant, and the noise with Gaussian distribution in amplitude and constant power spectral density is superimposed.

In the following, the impact of PIM interference on the demodulation link under constant SIR will be discussed in two cases. One is the impact of PIM interference on the demodulation link when the center frequency of PIM interference falls exactly at the carrier frequency of the uplink received signal, and the phase is the same as the phase of the uplink received signal. In this case, PIM interference has the greatest impact on the performance of the uplink received signal and the worst BER. The other is the analysis of the impact of PIM interference on the demodulation link when the center frequency of PIM interference does not fall at the carrier frequency of the uplink received signal.

1 Analysis of the Impact of PIM Interference When the Center Frequency Is the Same as the Carrier Frequency of the Uplink Received Signal
Since any memoryless nonlinear behavior can be described by the power series model, the power series model is used to analyze the influence of PIM interference on demodulation

$$Y(t) = a_1 X(t) + a_2 X^2(t) + a_3 X^3(t) + \cdots + a_n X^n(t) + \cdots \tag{6.3}$$

where $Y(t)$ is the PIM interference signal; $X(t)$ is the downlink m-way carrier signal, which can be expressed as

$$X(t) = \sum_{i=1}^{m} R_i(t)\cos(2\pi f_i t) \tag{6.4}$$

where f_i is the frequency of the downlink carrier signal; $R_i(t)$ is the baseband shaping signal with downlink frequency f_i.

Due to the design characteristics of the upstream and downstream signal bands, only the PIM products in the region I (i.e., PIM products with PIM frequencies lower than the second harmonic frequency) are usually considered, so only the PIM interference generated by odd powers of $X(t)$ is considered.

Since the PIM product grows nonlinearly with the number of downlink carriers, only the PIM product of the downlink with dual-carrier and BPSK modulation is considered next. The PIM products of the downlink signal with multicarrier and other modulation modes can also be analyzed using this method, but it is more complicated.

The downlink $X(t)$ can be expressed as

$$X(t) = \sum_{i=1}^{2} R_i(t) \cos(2\pi f_i t) \qquad (6.5)$$

Then the $n(n=2p+1)$ order PIM product that falls into the receiving band can be expressed as

$$Y_n(t) = R_1^{p+1} R_2^p \cos\left(2\pi\left((p+1)f_1 - pf_2\right)\right) \qquad (6.6)$$

When the baseband signal is modulated using rectangular pulse BPSK (binary phase shift keying), $R_1=\pm1$ and $R_2=\pm1$, so $R_1^{p+1} R_2^p = \pm1$. If R_1 and R_2 take the value of ±1 with equal probability, then the values of $R_1^{p+1} R_2^p$ are also taken with equal probability.

When the uplink signal frequency $f_{up}=(p+1)f_1-pf_2$, then the PIM interference has the greatest impact on the reception of the uplink signal. The PIM interference signal after down-conversion to get the baseband signal can be expressed as

$$Y_b(t) = R_1^{p+1} R_2^p = \pm1 \qquad (6.7)$$

At the uplink signal receiving end, the signal entering the adjudicator can be expressed as

$$z = \sqrt{P_s} X + \sqrt{P_I} Y_b + n \qquad (6.8)$$

where X is the uplink signal per unit energy, Y_b is the interference signal per unit energy, n is the additive Gaussian white noise, P_s is the signal power, P_I is the interference power.

The BER is calculated for $Y_b=+1$ and $Y_b=-1$, respectively. The BER can be expressed as

$$P_e = P(Y_b = +1)P(X = +1)P(z < 0 | Y_b = +1, X = +1)$$

$$+ P(Y_b = -1)P(X = +1)P(z < 0 | Y_b = -1, X = +1)$$

$$+ P(Y_b = +1)P(X = -1)P(z > 0 | Y_b = +1, X = -1)$$

$$+ P(Y_b = -1)P(X = -1)P(z > 0 | Y_b = -1, X = -1) \qquad (6.9)$$

When $X=\pm1$ with equal probability and $Y_b=\pm1$ with equal probability, Equation 6.9 can be expressed as

$$P_e = \frac{1}{4}P(z < 0 | Y_b = +1, X = +1) + \frac{1}{4}P(z < 0 | Y_b = -1, X = +1)$$

$$+ \frac{1}{4}P(z > 0 | Y_b = +1, X = -1) + \frac{1}{4}P(z > 0 | Y_b = -1, X = -1) \qquad (6.10)$$

The statistical distribution of the signals entering the adjudicator is considered below. Four probability density functions for the conditions $X=\pm 1$ and $Y_b=\pm 1$ can be written as

$$f_{X=+1,Y_h=+1}(z)=\frac{1}{\sqrt{2\pi}\sigma}\exp\left(-\frac{\left(z-\left(\sqrt{P_s}+\sqrt{P_I}\right)\right)^2}{2\sigma^2}\right) \tag{6.11}$$

$$f_{X=+1,Y_h=-1}(z)=\frac{1}{\sqrt{2\pi}\sigma}\exp\left(-\frac{\left(z-\left(\sqrt{P_s}-\sqrt{P_I}\right)\right)^2}{2\sigma^2}\right) \tag{6.12}$$

$$f_{X=-1,Y_I=+1}(z)=\frac{1}{\sqrt{2\pi}\sigma}\exp\left(-\frac{\left(z-\left(-\sqrt{P_s}+\sqrt{P_I}\right)\right)^2}{2\sigma^2}\right) \tag{6.13}$$

$$f_{X=-1,Y_h=-1}(z)=\frac{1}{\sqrt{2\pi}\sigma}\exp\left(-\frac{\left(z-\left(-\sqrt{P_s}-\sqrt{P_I}\right)\right)^2}{2\sigma^2}\right) \tag{6.14}$$

Substituting equations (6.11)–(6.14) into equation (6.10), the uplink received BER of the system can be expressed as

$$P_e=\frac{1}{4}\int_{-\infty}^{0}f_{X=+1,Y_h=+1}(z)dz+\frac{1}{4}\int_{-\infty}^{0}f_{X=+1,Y_b=-1}(z)dz$$
$$+\frac{1}{4}\int_{0}^{+\infty}f_{X=-1,Y_h=+1}(z)dz+\frac{1}{4}\int_{0}^{+\infty}f_{X=-1,Y_b=-1}(z)dz \tag{6.15}$$

By simplifying equation (6.15), it can be expressed as

$$P_e=\frac{1}{2}\int_{-\infty}^{0}f_{X=+1,Y_I=+1}(z)dz+\frac{1}{2}\int_{-\infty}^{0}f_{X=+1,Y_b=-1}(z)dz$$
$$=\frac{1}{4}\mathrm{erfc}\left(\left(1+\frac{1}{\sqrt{SIR}}\right)\sqrt{SNR}\right)+\frac{1}{4}\mathrm{erfc}\left(\left(1-\frac{1}{\sqrt{SIR}}\right)\sqrt{SNR}\right) \tag{6.16}$$

where SIR is the signal-to-interference ratio, $SIR=P_s/P_I$, P_s is the signal power; SNR is signal to noise ratio, $SNR=P_s/N_0$, and N_0 is the noise power, $N_0=2\sigma^2$.

The analysis and simulation curves of BER with SIR for uplink communication receivers under SNR=0 dB and SNR=5 dB are given in Figure 6.3.

From Figure 6.3, it can be seen that the analytical curve of the BER variation with the SIR is consistent with the general trend of the simulation curve. The simulation curve basically matches the analysis curve, which indicates that the analysis model is basically reasonable.

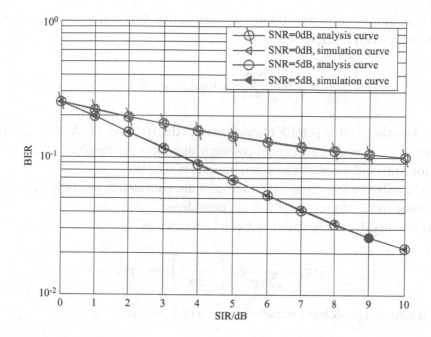

FIGURE 6.3 Analysis and simulation curves of BER variation with signal-to-interference ratio at constant SIR.

The above analysis is based on the assumption that the PIM interference power is constant, the center frequency of the PIM interference is exactly equal to the center frequency of the uplink received signal and the phase of the PIM interference is consistent with the phase of the uplink received signal. The impact of PIM interference on the uplink received signal is the largest, i.e., the above analysis gives the worst performance of the uplink received signal demodulation performance with constant SIR.

2 Analysis of the Impact of the Center Frequency of PIM Interference with Different Carrier Frequencies of Uplink Received Signals

The previous section introduced the analysis on the impacted demodulation performance of the uplink receiver in the case where the center frequency of PIM interference is the same as the carrier frequency of the uplink received signal. However, in the actual communication system, the center frequency of PIM interference is often different from the uplink received signal carrier frequency. Next, the analysis of the demodulation performance in the case where the center frequency of PIM interference is different from the carrier frequency of the uplink received signal is introduced.

Similar to the previous case, since the power series model can approximate any continuous memoryless nonlinear behavior, the power series model can be used to describe the nonlinear characteristics of passive microwave components, i.e.

$$Y(t) = a_1 X(t) + a_2 X^2(t) + a_3 X^3(t) + \cdots + a_n X^n(t) + \cdots \qquad (6.17)$$

where n is the order; $Y(t)$ is the PIM interference signal; $X(t)$ is m-way carrier signal in the downlink, which can be expressed as

$$X(t) = \sum_{i=1}^{m} R_i(t)\cos\left(2\pi f_i t + \theta_i\right) \tag{6.18}$$

where the signal sampling point X obeys Gaussian distribution, i.e., $X \sim N(0, \sigma^2)$; the envelope R_i obeys independent and equal Rayleigh distribution; the phase θ_i obeys uniform distribution with independent and equal interval of $(0, 2\pi)$.

Similarly, only the PIM products in region I are considered, i.e., only the PIM disturbances arising from odd powers of $X(t)$ are considered.

The probability density function of X can be written as

$$f_X(x) = \frac{1}{\sqrt{2\pi}\sigma}\exp\left(-\frac{x^2}{2\sigma^2}\right), -\infty < x < +\infty \tag{6.19}$$

When n is odd, the probability density function of $Y_n = X^n$ is

$$f_Y(y) = \frac{1}{\sqrt{2\pi}\sigma}\exp\left(-\frac{y^{2/n}}{2\sigma^2}\right) \cdot \frac{1}{k} \cdot y^{-(n-1)/n} \tag{6.20}$$

It is more complicated to analyze its probability density directly based on the power series model, but comparing it with the classical probability distribution, it is found to be approximately similar to the Cauchy distribution, as shown in Figure 6.4.

The probability density function of the Cauchy distribution can be written as

$$f(x) = \frac{1}{\pi}\frac{\lambda}{x^2 + \lambda^2} \tag{6.21}$$

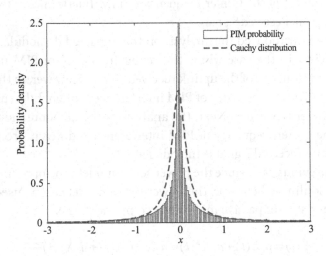

FIGURE 6.4 The 9th order PIM probability distribution vs. the Cauchy distribution.

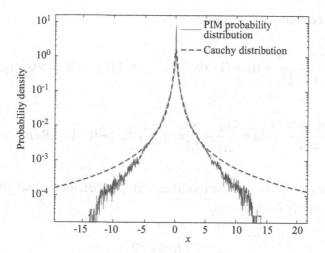

FIGURE 6.5 The 9th order PIM probability distribution vs. the Cauchy distribution plotted in logarithmic form.

where λ is the scale parameter, which controls the concentration degree of the probability distribution.

As can be seen from the figure, both roughly match, which indicates that the probabilistic statistical properties of PIM interference are basically reasonable under the estimation of the Cauchy distribution.

The peak, mode and median of the above probability distribution are 0. However, the mean, variance and moments of the Cauchy distribution are not defined, which is not conducive to subsequent analysis. As shown in Figure 6.5, the probability density curves are plotted in logarithmic form, which can be seen to be more different on the sides away from the median, and the tails of the Cauchy distribution are heavier.

Therefore, if one wants to approximate the PIM probability density using the Cauchy distribution, one needs to correct the Cauchy distribution by making its tails lighter. In contrast, the Gaussian distribution as a light-tailed distribution can help to correct the Cauchy distribution. The corrected probability density can be expressed as

$$f(x) = \frac{1}{C} \cdot \frac{\lambda}{x^2 + \lambda^2} \cdot \exp\left(\frac{-x^2}{K^2}\right) \tag{6.22}$$

where

$$C = \pi\left(1 - \operatorname{erf}\left(\frac{\lambda}{K}\right)\right) \cdot \exp\left(\frac{\lambda}{K}\right)^2 \tag{6.23}$$

where K is Gaussian correction parameter.

After curve fitting, the relationship between κ and n and the relationship between λ and n are obtained:

$$\lambda = 0.62 \times e^{-0.13n} \tag{6.24}$$

$$K = 2.6 \times e^{0.12n} \tag{6.25}$$

According to the reference [2], it is known that

$$\int_0^{+\infty} \frac{\exp(-\mu^2 x^2)}{x^2+\beta^2}\,dx = (1-\Phi(\beta\mu))\frac{\pi}{2\beta}\exp(\beta^2\mu^2), \mathrm{Re}\,\beta>0, |\arg\mu|<\frac{\pi}{4} \quad (6.26)$$

$$\int_0^{+\infty} \frac{x^2\exp(-\mu^2 x^2)}{x^2+\beta^2}\,dx = \frac{\sqrt{\pi}}{2\mu}-\frac{\pi}{2\beta}\exp(\beta^2\mu^2)(1-\Phi(\beta\mu)), \mathrm{Re}\,\beta>0, |\arg\mu|<\frac{\pi}{4} \quad (6.27)$$

As a result, it is easy to verify that the modified Cauchy distribution satisfies the basic properties of the probability distribution.

$$\begin{cases} f_n(x) \geq 0 \\ F_n(\infty) = \int_{-\infty}^{+\infty} f_n(x)dx = \frac{1}{C}\int_{-\infty}^{+\infty}\exp\left(\frac{-x^2}{K^2}\right)\frac{\lambda}{x^2+\lambda^2}\,dx = 1 \end{cases} \quad (6.28)$$

where $f_n(x)$ is the modified Cauchy distribution probability density; $F_n(\infty)$ represents that the integral of $f_n(x)$ at $(-\infty, +\infty)$ is 1.

Existence expectation and variance are

$$E_n(x) = \int_{-\infty}^{+\infty} x \cdot f_n(x)dx = \frac{1}{A}\int_{-\infty}^{+\infty}\exp\left(\frac{-x^2}{K^2}\right)\frac{\lambda x}{x^2+\lambda^2}\,dx = 0 \quad (6.29)$$

$$D_n(X) = \frac{1}{C}\int_{-\infty}^{+\infty}\exp\left(\frac{-x^2}{K^2}\right)\frac{\lambda x^2}{x^2+\lambda^2}\,dx = \frac{\lambda K}{\sqrt{\pi}\left(1-\mathrm{erf}\left(\frac{\lambda}{K}\right)\right)\cdot\exp\left(\left(\frac{\lambda}{K}\right)^2\right)} - \lambda^2 \quad (6.30)$$

3 BER Analysis of Multicarrier PIM Interference on Communication Receivers
Considering the nonnegligible effect of noise in the system, the system interference after adding noise is $Z(t) = y(t) + n(t)$, where $y(t)$ is the PIM interference, and $n(t)$ is the receiver noise, whose probability density $g(x)$ is usually represented by the Gaussian distribution $N(0, \sigma^2)$ with mean value 0, i.e.

$$g(x) = \frac{1}{\sqrt{2\pi}\sigma}\exp\left(\frac{-x^2}{2\sigma^2}\right) \quad (6.31),$$

where σ is the standard deviation.

If only a single-order PIM signal is considered, the probability density of $y(t)$ can be expressed by equation (6.31). $y(t)$ and $n(t)$ are independent of each other, so the probability density function of $Z(t)$ is the convolution of the probability density $f(*)$ of $y(t)$ with the probability density $g(x)$ of $n(t)$.

$$f_Z(s) = f(x) * g(x) = \int_{-\infty}^{+\infty} f(t) \cdot g(s-t)dt$$

$$= \int_{-\infty}^{+\infty} \frac{1}{C} \cdot \frac{\lambda \cdot \exp\left(\dfrac{-t^2}{K^2}\right)}{t^2 + \lambda^2} \cdot \exp\left(\dfrac{-(s-t)^2}{\sigma^2}\right)dt$$

$$= \pi \cdot \frac{\mathrm{Re}\left(\exp\left(-\left(\dfrac{s+i\beta}{\alpha}\right)\right)\left(1 + \dfrac{2i}{\sqrt{\pi}}\int_0^{\frac{4+\beta}{\alpha}}\exp\left(t^2\right)dt\right)\right)\exp\left(\dfrac{-z^2}{\gamma}\right)}{C \cdot \sqrt{2\pi}\sigma} \qquad (6.32)$$

where

$$\alpha = \frac{\sqrt{2}\sigma\sqrt{2\sigma^2 + K^2}}{K} \qquad (6.33)$$

$$\beta = \lambda \cdot \frac{2\sigma^2 + K^2}{K^2} \qquad (6.34)$$

$$\gamma = 2\sigma^2 + K^2 \qquad (6.35)$$

Define the interference-to-noise ratio (INR) as

$$\mathrm{INR} = \frac{P_{\mathrm{pim}}}{P_{\mathrm{noise}}} \qquad (6.36)$$

where P_{pim} is the power of PIM; P_{noise} is the noise power.

For a given value of INR, the parameter σ can be calculated from the following equation.

$$\sigma = \left(\left(\frac{\lambda K}{\sqrt{\pi}\left(1 - \mathrm{erf}\left(\dfrac{\lambda}{K}\right)\right) \cdot \exp\left(\left(\dfrac{\lambda}{K}\right)^2\right)} - \lambda^2\right)\mathrm{INR}^{-1}\right)^{\frac{1}{2}} \qquad (6.37)$$

Comparing the probability density function obtained from the theoretical derivation with the probability density curve of the measured PIM and the simulated PIM probability density curve, we can get the results shown in Figure 6.6. From the figure, it can be seen that the three can be in good agreement.

The following is an example of BPSK modulation mode to introduce the impact of multicarrier PIM interference on the receiver BER. Let the probability of sending code "1" and

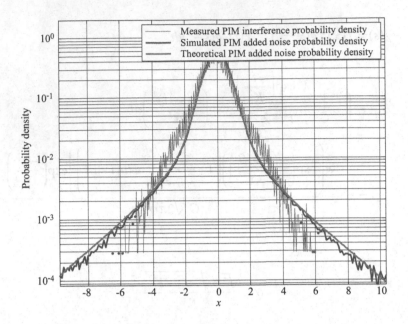

FIGURE 6.6   Comparison of probability density after noise addition.

code "0" is $P(1)$ and $P(0)$, respectively, and the transmission probability is equal, that is, $P(0)=P(1)=1/2$, and let $P(1/0)$ and $P(0/1)$ be the probability of sending "0" receive "1" and send "1" receive "0" error probability, the system's BER can be expressed as

$$P = P(0)P(1/0)+P(1)P(0/1)$$

$$= \frac{1}{2}P(1/0)+\frac{1}{2}P(0/1)$$

$$= \frac{1}{2}\int_{-\infty}^{0} f_Z(s-A)dz+\frac{1}{2}\int_{0}^{+\infty} f_Z(s+A)ds$$

$$= \int_{A}^{+\infty} f_Z(s)ds \tag{6.38}$$

Therefore, the system BER is determined by the probability density of $Z(t)$. Substituting Equation 6.32 into equation (6.38), the system BER under single-order PIM interference can be obtained as

$$P = \int_{A}^{+\infty} \pi \cdot \frac{\mathrm{Re}\left(\exp\left(-\left(\frac{s+i\beta}{\alpha}\right)\right)\left(1+\frac{2i}{\sqrt{\pi}}\int_{0}^{\frac{5+i\beta}{\alpha}}\exp\left(t^2\right)dt\right)\right)\exp\left(\frac{-x^2}{\gamma}\right)}{C\cdot\sqrt{2\pi}\sigma} dx \tag{6.39}$$

According to the comparison of the results of theoretical calculation, simulation results, and actual measurement results as shown in Figure 6.7, it can be seen that the probability density model fits very well.

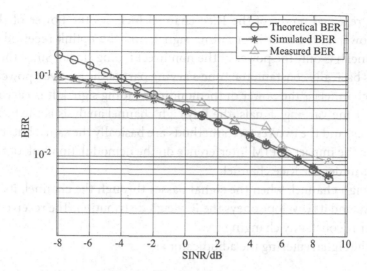

FIGURE 6.7 Comparison of the variation curve of BER with signal-to-noise ratio under PIM interference.

Figure 6.7 shows the curves of the BER of the receiver with the signal-to-noise ratio under the influence of the 9th-order PIM interference. Among them, the red line is the theoretical BER curve calculated by the derivation of equation (6.39), the blue line is the simulated BER curve in the communication system built in MATLAB®, and the green line is the measured BER curve. From Figure 6.10, it can be seen that the trend of the theoretical and simulated BER curves of the receiver with the variation of the SIR under the influence of PIM interference is basically the same. Under the same BER conditions, the errors of the theoretical and simulated curves are basically within 2 dB, and the errors of the measured results with the simulated and theoretical results are also within 2 dB, which shows the rationality of the analysis model.

The analysis of the influence of PIM interference on the demodulation link at constant SIR is more suitable for analyzing the influence of PIM interference on the demodulation performance in the burst communication. This is mainly because the communication time of burst communication is short, and the PIM interference power in a short period of time can be approximately regarded as constant. The method can be used to analyze the BER of uplink communication link under PIM interference at a certain constant SIR.

6.2.2.2 Analysis of the Effect of PIM Interference on Demodulation Process at Time-Varying SIR

The previous section focused on the effect of PIM interference on the demodulation link at constant SIR, but PIM interference is time-varying and with a constantly changing SIR. This section will focus on the analysis of the impact of PIM interference on the demodulation link at the time-varying SIR. The general idea is to analyze the statistical characteristics of the time-varying SIR, and then find out the variation of the average BER with the average SIR for that period of time.

The uplink received power of the Rice channel includes the power of the direct signal and the power of the nondirect fading signal, and the uplink received power of the Rayleigh channel has only the power of the nondirect fading signal. Since the power of the direct signal is basically constant, the time-varying part of the received power of both Rice and Rayleigh channels is the power of the nondirect fading signal. It is easy to see that, in the analysis of time-varying signals, the Rayleigh channel analysis is the basis of the Rice channel analysis, and the two research methods are basically the same, but it is also necessary to analyze the impact of PIM interference on the demodulation link of the communication receiver under the Rice channel.

In the Rayleigh channel, when the signal passes through the channel, its signal amplitude is random, and its envelope obeys the Rayleigh distribution. The received power of the uplink receiver is constantly changing.

Represent the signal entering the adjudicator as

$$Z = \sqrt{P_s} h_0 x + \sqrt{P_I} h_1 y + n \tag{6.40}$$

where x is uplink received signal per unit energy; y is interference signal per unit energy; n is AWGN (Additive White Gaussian Noise); P_s is average power of uplink received signal; P_I is average power of PIM interference; h_0 and h_1 are attenuation coefficients of the changing uplink received signal and PIM interference. In the following analysis, it is assumed that h_0 and h_1 obey Rayleigh distribution.

After entering the judgment, the SINR (Signal to Interference plus Noise Ratio) can be expressed as

$$\text{SINR} = \frac{X}{Y + \sigma^2} \tag{6.41}$$

where $X = |h_0|2P_s$, $Y = |h_1|2P_I$, X and Y obey exponential distribution. Their probability density functions are denoted as

$$f_X(X) = \frac{1}{P_s} \exp\left(-\frac{X}{P_s}\right) \tag{6.42}$$

$$f_Y(Y) = \frac{1}{P_I} \exp\left(-\frac{Y}{P_I}\right) \tag{6.43}$$

The statistical properties of the SINR are discussed below. The cumulative distribution function of the SINR can be expressed as

$$F_{\text{SINR}}(\gamma) = \Pr(\text{SINR} \leq \gamma) = \int_0^{+\infty} \Pr\left(X \leq \gamma\left(Y + \sigma^2\right)\right) f_Y(Y) dY \tag{6.44}$$

Substituting Equations 6.42 and 6.43 into Equation 6.44, Equation 6.44 can be simplified as

$$F_{\text{SINR}}(\gamma) = 1 - \frac{P_s}{P_s + \gamma P_I} \exp\left(-\frac{\gamma \sigma^2}{P_s}\right) \tag{6.45}$$

In the case of time-varying SINR, the effect of PIM interference on the receiver BER can be expressed as

$$\text{SER}_L = \int_0^{+\infty} a_{\text{mod}} Q\left(\sqrt{2b_{\text{mod}}\gamma}\right) f_{sINR}(\gamma) d\gamma \qquad (6.46)$$

where $Q(\cdot)$ is Gaussian Q function; a_{mod} and b_{mod} are parameters related to the modulation mode.

Substituting the Q function into Equation 6.46, Equation 6.46 can also be expressed as

$$\text{SER}_L = \frac{a_{\text{mod}}\sqrt{b_{\text{mod}}}}{2\sqrt{\pi}} \int_0^{+\infty} \frac{\exp(-b_{\text{mod}}\gamma)}{\gamma^{1/2}} F_{\text{SINR}}(\gamma) d\gamma \qquad (6.47)$$

Under BPSK modulation, $a_{\text{mod}}=1$ and $b_{\text{mod}}=1$. Substituting Equation 6.46 into Equation 6.47, the relation between BER, average SIR and average SNR under BPSK modulation can be obtained after calculation as

$$P_e = \frac{1}{2} - \frac{1}{2}\sqrt{SIR} \cdot \exp(SIR(1+1/(2SNR)))\Gamma\left(\frac{1}{2}, SIR(1+1/(2SNR))\right) \qquad (6.48)$$

where $SIR = P_s/P_I$; $SNR = P_s/N_0$; $N_0 = 2\sigma^2$; and $\Gamma(\alpha, x)$ is defined as follows:

$$\Gamma(a,x) = \int_0^{+\infty} e^{-t} t^{a-1} dt \qquad (6.49)$$

Figure 6.8 gives the analytical and simulation curves of the BER variation with the average SIR for the uplink with SNR=5 dB and SNR=25 dB under PIM interference.

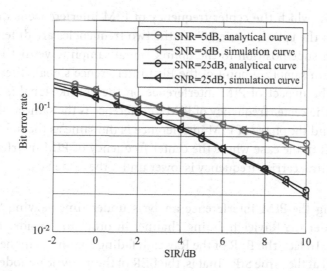

FIGURE 6.8 Analytical and simulation curves of BER variation with average SIR under time-varying SIR.

From Figure 6.8, it can be seen that the analytical curve of BER variation with SIR is consistent with the general trend of the simulation curve. The error between the simulation curve and the analysis curve is within 2 dB, which shows that the analysis of the effect of PIM interference on the demodulation performance of the communication receiver under the time-varying SIR is basically reasonable.

The above analysis is analyzed on basis that the uplink received power and PIM interference power are subject to exponential distribution; if the uplink received power and PIM interference power are subject to other statistical characteristics, then the analysis with the above ideas can also get more satisfactory results.

The model of the effect of PIM interference on the demodulation performance of communication receivers under time-varying SIR is more suitable for analyzing the effect of PIM interference on the demodulation performance of communication systems for long-time continuous communication than the model of the effect of PIM interference on the demodulation performance of communication receivers under constant SIR. This is mainly because the communication time of long-time continuous communication is long, and its PIM interference power is constantly changing, and the longer the time, the more its SIR distribution approximately obeys its theoretical SIR statistical characteristics, and the method can be used to analyze the BER of PIM interference on uplink communication link under time-varying SIR.

The effect of PIM interference on receiver demodulation performance at a constant SIR is analyzed in two models.

1. The model in which the center frequency of PIM interference is the same as the frequency of the uplink carrier and the phase of PIM interference is the same as the phase of the uplink carrier. This is the case where the PIM interference has the greatest impact on the uplink signal at a constant SIR.

2. The model in which the center frequency of PIM interference is different from the frequency of the uplink carrier. Since the two frequencies are different, the interference at each superposition to the uplink signal sampling verdict is not necessarily the maximum sampling point of the PIM interference signal. Therefore, under this condition, the impact of PIM interference on the uplink signal is smaller than the case where the center frequency of PIM interference is the same as the uplink carrier frequency and the phase of PIM interference is the same as the uplink carrier phase, and the BER of the case where the center frequency of PIM interference is different from the uplink carrier frequency is lower under the same SIR.

When performing the PIM interference analysis under time-varying SIR, the analysis is performed under the Rayleigh fading channel in order to consider the time-varying nature of the SIR. In fact, the BER of the Rayleigh fading channel is higher than that of the Gaussian channel at the same SIR. That is, the BER of the analytical model under the time-varying SIR is higher when the SIR in the constant SIR analysis is the same as the average SIR in the time-varying SIR analysis.

6.3 LOW-PIM DESIGN PRINCIPLES

6.3.1 System Solution Design

The first four of the five mobile communications satellite systems deployed by the United States in the 1970s and 1980s have encountered PIM problems. The first three (FLTSATCOM, MARISAT and MARECS) encountered problems during the system integration and testing phases of the program, resulting in delays of 1 or 2 years. In addition to the delays, program costs increased due to the rework and retesting involved. In the fourth satellite system, INTELSAT V MCS, there are no PIM problems during the ground test phase, but the system occasionally experienced random noise bursts due to in-orbit PIM generation.

If common paths for low and high power signals cannot be avoided, then proper selection of transmit and receiving frequencies is the starting point for reducing PIM noise, and transmit and receiving bands should be separated as widely as possible in the frequency range. The reverse example is the U.S. FLTSATCOM satellite system. In current multi-channel systems, it is difficult to achieve complete frequency separation, but the best way to minimize PIM interference is to separate the two bands as much as possible.

6.3.2 Avoid Ferromagnetic Materials

An important technique to verify that PIM generation is caused by ferromagnetic materials is described by Young of the U.S. Naval Laboratory in reference [4]. PIM-generating components placed in an externally generated DC magnetic field can reduce PIM by 10–20 dB, if the magnetic field is applied perpendicular to the component axis, and the magnetic field is rotated parallel to the connector axis, it further reduces PIM by about 10 dB, for a total reduction of 20–30 dB compared to the initial value.

Typical results obtained in these studies are shown in Figure 6.9, which indicates the influence of different metal materials on PIM level. Therefore, ferromagnetic materials and corrosive materials are the main causes of PIM.

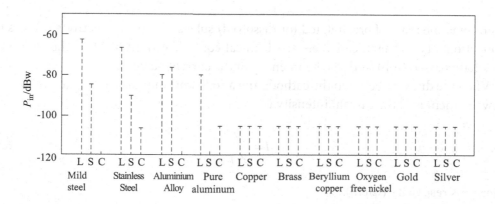

FIGURE 6.9   PIM values at 1.5 GHz, 3 GHz, 6 GHz near the center frequency P_{in}. (Note: L, S and C indicate the center frequencies around 1.5GUz, 3GUz and 6GHz, respectively; the total incident power is 3.2W, and the axial force is 10.65N.)

6.3.3 Waveguide Connection

Carlos studied the PIM level values of the waveguide flange at different connection pressures. It is found that the PIM level decreases slowly with increasing applied pressure and continues to a certain pressure from which the PIM level drops sharply. This is because at relatively low applied pressures, the contact resistance is mainly determined by the non-contact capacitance, and it is directly related to the surface and the contact area is very small, the PIM value decreases slowly, and the change is not too large for different applied pressures. Since these resistances depend linearly on the contact area, the contact capacitance and shrinkage resistance provide the electrical response of the junction as the pressure increases, causing the PIM level to decrease more rapidly.

6.3.4 Surface Treatment

Surface treatment is an effective means to improve material nonlinearity. Microwave passive components are generally made of aluminum, which has poor PIM characteristics, and if it is not surface treated, the components made of it will produce high PIM levels. Silver has low PIM characteristics, but it is expensive and should not be used directly in the manufacture of components, so the use of electroplating on the surface of aluminum products is a better way.

6.4 SURFACE TREATMENT METHOD FOR LOW-PIM

6.4.1 Effect of Plating Process on PIM

Poor plating uniformity is one of the main problems in the current plating process. The distribution of current or current density on the surface of the plated part directly affects the uniformity of the thickness of the plating on the surface of the part. According to Faraday's first law, the mass of precipitated (or dissolved) material on the electrode during electrolysis is proportional to the amount of electricity passed.

$$m = k \cdot Q = k \cdot I \cdot t \tag{6.50}$$

where m is the mass of precipitated (or dissolved) substance on the electrode (g), k is the proportionality constant, called electrochemical equivalent [g/(A·h)], Q is the amount of electricity passed (A·h), and I is the intensity of the current passed (A).

When the distance between the cathode and anode working surfaces is y and the voltage between them is V, the current intensity is

$$I = \frac{V}{y \cdot \rho} \tag{6.51}$$

where ρ is resistivity (Ω/cm).

The plating thickness δ is

$$\delta = \frac{100 \cdot I \cdot t \cdot k \cdot \eta}{r \cdot s} \tag{6.52}$$

where I is current intensity (A); t is plating time (h); k is electrochemical equivalent [g/(A–h)]; η is current efficiency, and r is density of precipitated plated metal (g/cm³).

In theory, because the distance between the cathode and anode is much larger than the coating thickness, the current at each location during the plating process will remain constant. However, in the actual production process, affected by the "edge effect" and "deep cavity effect," the current on the internal and external working surfaces and different locations are not the same, which leads to a large difference in the thickness of the coating on each part, especially for the high-power microwave components with complex structure.

In terms of plating thickness and uniformity improvement, the electric field in the plating process of microwave passive parts (cavities) can be simulated and analyzed by using numerical electromagnetic field calculation simulation software. First, the distribution law of current or current density in each part of the inner cavity surface of the (complex) parts is simulated visually; then, several corrections are made to obtain the model with the most even current distribution; finally, the auxiliary anode or auxiliary cathode is produced according to the corrected model and is applied to the actual production process to make the current distribution on the surface of the plated parts more uniform. For cavity parts, dual power supply plating technology (i.e., the plating process of the inner cavity surface of the part is controlled by one plating power supply alone, and the outer surface of the part is controlled by another power supply) is used to control the inner and outer electric fields independently, so as to precisely control the plating thickness of the inner and outer cavity parts and improve the uniformity of the plating. Next, the surface morphology and microstructure of each group of plated layers are observed by SEM, and the thickness of the plated layers is examined by X-ray to optimize the current application method for thick silver plating and to determine the appropriate process parameters.

In order to better observe the PIM phenomenon generated by different plating materials on the internal connection surface of passive devices, a passive assembly is proposed to be designed, by which the PIM phenomenon is observed and studied by replacing the coaxial resonant rods with different plating layers. The coaxial resonant rod is processed separately and fastened to the cavity by fastening screws. The current at the root of the coaxial resonant rod is the largest, and the PIM phenomenon is most likely to be generated due to the nonlinearity. A passive 2nd-order coaxial filter is designed to generate dual-frequency multicarrier signals and avoid PIM from other parts inside the passive assembly affecting the system observation, as shown in Figure 6.10.

Using the filter cavity and the coaxial resonant rod, the surface of the coaxial resonant rod is prepared into different states of silver plating layer, and the PIM test of the sample is used to systematically analyze and study the influence of different plating layer thickness, surface roughness, hardness and wear resistance on the PIM to obtain the actual data.

When the plating thickness is less than the skinning depth, it has a greater effect on the PIM of the filter; when the plating thickness reaches more than two skinning depths, it has almost no effect on the PIM. The effect of different plating thicknesses on PIM is shown in Figures 6.11 and 6.12.

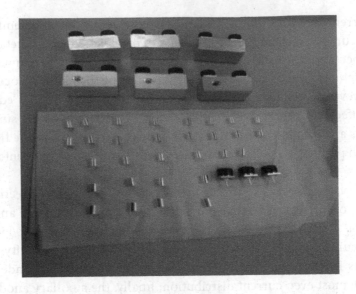

FIGURE 6.10   Surface treatment of the test specimens.

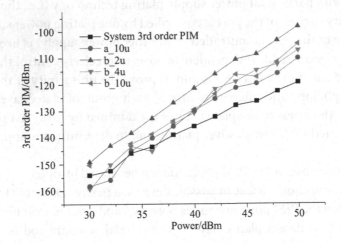

FIGURE 6.11   Effect of different plating thicknesses on PIM (3rd Order PIM).

6.4.2 Effect of Plating Process on Electrical Properties of Cavity Filters

In microwave circuits, resonant cavities are widely used in various microwave components, including filters, oscillators, frequency meters, tunable amplifiers, etc. Therefore, it is important to study the properties of resonant cavities. In the study of the properties of resonant cavities, the two most important parameters of resonant cavities are resonant frequency and quality factor. For an ideal resonant cavity, the resonant frequency is the frequency at which the average stored magnetic energy and electrical energy are equal, there is no loss in the resonant cavity, and the quality factor is infinite. However, the actual resonant cavity has losses. The quality factor of a resonant cavity is a measure of the loss of the resonant cavity, and the larger the loss, the lower the quality factor.

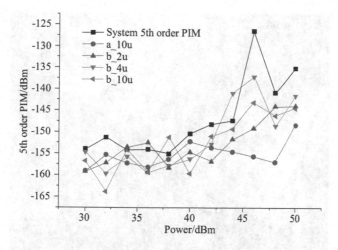

FIGURE 6.12   Effect of different plating thicknesses on PIM (5th Order PIM).

Conductor loss is one of the important parts of resonant cavity loss, including the loss caused by the limited conductivity of metal and the additional loss caused by the roughness of the metal surface. Without considering the additional loss caused by surface roughness, the resonant cavity conductor loss of some regular structures can be approximated by perturbation method, while for some complex structures, it can be modeled and solved by various numerical methods, but the conductor loss thus found may be problematic. This is because, as the operating frequency increases, the additional loss caused by surface roughness may be very large, which can seriously affect the quality factor of the resonant cavity. Therefore, the additional losses caused by surface roughness cannot be ignored when analyzing the quality factor of the resonant cavity.

6.4.2.1 Resonant Frequency and Quality Factor of Rough Lossy Cavity for Eigenmode Analysis

In the eigenmode analysis, the cavity is treated as a closed cavity without the coupling of ports and external circuits, and the field and resonant frequency in the resonant cavity can be found by Maxwell's equations with boundary conditions. When the resonant cavity has losses, the resonant frequency is a complex number, whose real part indicates the resonant frequency, and imaginary part indicates the losses, and the quality factor Q can be expressed as

$$Q = \frac{f_R}{2f_I} \quad (6.53)$$

where f_R is the real part of resonant frequency; f_I is the imaginary part of resonant frequency.

The resonant frequency and quality factor of the resonant cavity are calculated separately by the eigenmode analysis method below. The coaxial cavity model is shown in Figure 6.13.

FIGURE 6.13 Coaxial cavity model.

1. The cavity wall is considered as a smooth ideal conductor, i.e., the cavity wall is considered as a smooth ideal conductor without considering the surface roughness of the conductor and any loss. Its resonant frequency f_0 is

$$f_0 = \frac{c}{\lambda_0} = \frac{c}{2l} = 4.083 \,(\text{GHz}) \tag{6.54}$$

where l is the length of the resonant cavity.

2. The cavity wall is a smooth and lossy conductor, i.e., the surface is considered as smooth one, but the conductor loss of the coaxial cavity wall is considered. Using the complex frequency method, the quality factor Q of the transmission line resonant cavity can be obtained as

$$Q = \frac{\pi l}{\left(\alpha l + \dfrac{2R_s}{\eta}\right)\lambda_g} \left(\frac{\lambda_k}{\lambda}\right)^2 \tag{6.55}$$

where α is the attenuation constant of the transmission line, $\alpha = 2R_s\left(\dfrac{1}{a}+\dfrac{1}{b}\right)\bigg/\left(\eta \ln \dfrac{b}{a}\right)$; a and b are long and narrow sides of the waveguide, respectively; R_s is the surface impedance of the conductor; η is the wave impedance; λ_g is the waveguide wavelength. λ is free space wavelength. In the coaxial line, $\lambda = \lambda_g$.

The coaxial line parameters are substituted into the formula (6.52), and the Q value of the resonant cavity of the coaxial line can be obtained as

$$Q = \frac{\beta l}{2\alpha l + \dfrac{4R_s}{\eta}} \tag{6.56}$$

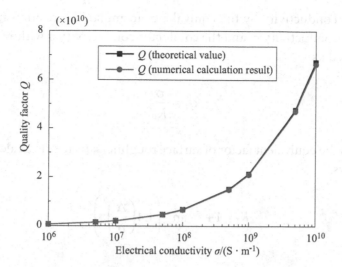

FIGURE 6.14   Variation of no-load quality factor Q with conductivity σ.

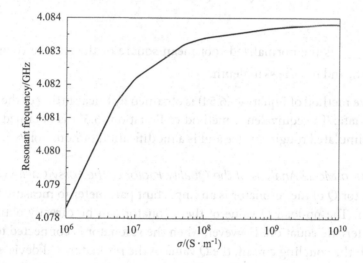

FIGURE 6.15   Variation of resonant frequency with electrical conductivity.

where $\beta = 2\pi/\lambda$.

When the conductor wall conductivity is $\sigma = 5.8 \times 107$ S/m, the Q value obtained is 5121.9, and the calculated resonant frequency is $f_0 = 4.0829$ GHz. Figure 6.14 shows the variation of quality factor with the cavity wall conductivity for the theoretical value and numerical calculation results, from which it can be seen that the resonant frequency with loss is slightly smaller than the resonant frequency with ideal conductor, and with the increase of conductivity, the resonant frequency gradually increases and finally converges to the resonant frequency with ideal conductor (Figure 6.15).

3. The cavity wall is a rough lossy conductor, i.e., both surface roughness and conductor are considered lossy. When the root-mean-square value of the roughness height is comparable to the skin depth, the effect of roughness can be equated to the

equivalent conductivity by the equivalence formula. The relationship between the equivalent conductivity σ_e and the conductor conductivity σ without roughness can be expressed as

$$\sigma_c = \frac{\sigma}{K_i^2} \tag{6.57}$$

where K_i is the equivalent factor of surface roughness; $i = 1, 2$. The calculation formula is as follows:

$$K_1 = 1 + \frac{2}{\pi} \tan^{-1}\left(1.4\left(\frac{\Delta}{\delta}\right)^2\right) \tag{6.58}$$

$$K_2 = 1 + \frac{2}{\pi} \tan^{-1}\left(\left(\frac{\Delta}{\delta}\right)^2\left(0.094\left(\frac{\Delta}{\delta}\right)^2 - 0.74\left(\frac{\Delta}{\delta}\right) + 1.87\right)\right) \tag{6.59}$$

where $\dfrac{\Delta}{\delta}$ —— is the normalized root mean square of the surface roughness for the skin depth, and δ is the skin depth.

The equivalence method of Equation (6.54) is obtained by linear fitting of the experimental measurement data. The equivalence method of Equation (6.55) is obtained by fitting the results of the simulated rough surface and is a modification of Equation 6.56.

6.4.2.2 Excitation Mode Analysis of the Quality Factor of the Lossy Cavity On-load

The quality factor Q of the resonator is an important parameter to measure the loss of the resonator itself. The no-load Q value of the resonator can be directly obtained by solving the characteristic equation. However, when the resonator is connected to the external circuit through the coupling circuit, the Q value of the resonator will deviate from the no-load Q value due to the influence of the external Q value. Generally, when calculating the on load Q value, we can get the following results through its relative bandwidth:

$$Q = \frac{1}{BW} = \frac{f_0}{\Delta f} \tag{6.60}$$

where BW is the relative bandwidth; f_0 is the center frequency; Δf is the absolute bandwidth.

It can be seen from Equation 6.57 that when the value of Q is relatively large, a very small change in Δf will cause a large error in Q. When using software simulation, the scattering parameters at discrete frequency points can only be obtained, but it is very difficult to get the accurate value of Δf. Therefore, when the Q value is high, it may bring large error by using the relative bandwidth to calculate the Q value.

The equivalent circuit of the resonator coupled to the external circuit through the coupling ring is shown in Figure 6.16.

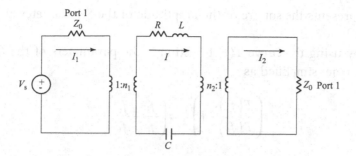

FIGURE 6.16   Equivalent circuit of resonant cavity coupled to external circuit.

When port 2 is connected with matching load, the maximum power absorbed by load is

$$P_L = I_2 I_2^* Z_0 = \frac{V_s^2}{Z_0} \frac{\beta_1 \beta_2}{\left(1 + \beta_1 + \beta_2\right)^2 + Q_0^2 \left(\dfrac{f}{f_0} - \dfrac{f_0}{f}\right)^2} \tag{6.61}$$

where

$$\beta_1 = \frac{n_1^2 Z_0}{R}, \beta_2 = \frac{n_2^2 Z_0}{R}, f_0 = \frac{1}{4\pi^2 LC}, \quad Q_0 = \frac{2\pi f_0 L}{R}.$$

When port 1 is matched, the maximum power that can be transmitted to the network is

$$P_{in} = I_1 I_1^* Z_0 = \frac{V_s^2}{4 Z_0} \tag{6.62}$$

Let $T(f) = \dfrac{P_L}{P_{in}}$, then $T(f) = \left|S_{21}(f)\right|^2$, and

$$T(f) = \frac{P_L}{P_{in}} = \frac{4\beta_1 \beta_2}{\left(1 + \beta_1 + \beta_2\right)^2 + Q_0^2 \left(\dfrac{f}{f_0} - \dfrac{f_0}{f}\right)^2} \tag{6.63}$$

where S21 (f) is the transmission characteristic of the network.

When $f = f_0$, $T(f_0) = \dfrac{4\beta_1 \beta_2}{\left(1 + \beta_1 + \beta_2\right)^2}$ and $Q_0 = Q(1 + \beta_1 + \beta_2)$, so when $\beta_1 \ll 1$ and $\beta_2 \lll 1$, we have

$$T(f) = \frac{T(f_0)}{1 + Q^2 \left(\dfrac{f}{f_0} - \dfrac{f_0}{f}\right)^2} \tag{6.64}$$

In the vicinity of f_0, when $f = f_k$, we have

$$T(f_k) = \frac{T(f_0)}{1 + Q^2 \left(\dfrac{f_k}{f_0} - \dfrac{f_0}{f_k}\right)^2} \tag{6.65}$$

where $T(f_k)$ represents the square of the amplitude of the S-parameter at f_k, i.e., $T(f_k) = |S21(f_k)|^2$.

Therefore, by using the curve $T(f)$ to estimate the parameters of the Q, the formula (6.63) can be further simplified as

$$\left(\frac{T(f_0)}{T(f_k)} - 1 \right)^{1/2} = \left(\frac{f_k}{f_0} - \frac{f_0}{f_k} \right) Q \tag{6.66}$$

Equation 6.63 is a linear function on the quality factor Q. Therefore, the parameters of Q can be estimated by the linear least square method. Calculate $|S21(f_k)|$ at equal intervals on both sides of f_0 for N points in total. Let $x_k = \frac{f_k}{f_0} - \frac{f_0}{f_k}$, $y_k = \left(\frac{T(f_0)}{T(f_k)} - 1 \right)^{1/2}$ and

$\bar{x} = \frac{1}{N} \sum_{k=1}^{N} x_k, \bar{y} = \frac{1}{N} \sum_{k=1}^{N} y_k$, then

$$\hat{Q} = \frac{\sum_{k=1}^{N}(x_k - \bar{x})(y_k - \bar{y})}{\sum_{k=1}^{N}(x_k - \bar{x})^2} \tag{6.67}$$

When $f_k = f_0 \pm \frac{1}{2}\Delta f_{3\,\text{dB}}$, $T(f_k) = \frac{1}{2}T(f_0)$, it can be obtained by formula (6.64) as $Q = \frac{f_0}{\Delta f_{3\,\text{dB}}}$.

Next, we use the calculation data to calculate the quality factor of coaxial cavity.

When the cavity wall is smooth ideal conductor, the external quality factor of the cavity is equal to the loaded quality factor. The calculated resonant frequency is 4.083 GHz, and the quality factor is 7652.9. When the cavity wall is a smooth lossy conductor, the loaded quality factor of the cavity is determined by the external quality factor and the loaded quality factor. When the conductivity of cavity wall is $\sigma=5.8 \times 10^7$ S/m, the resonant frequency is 4.0829 GHz, and the loaded quality factor is 2071.9. When the cavity wall is a rough lossy conductor, the conductivity is still $\sigma=5.8 \times 10^7$ S/m. Using equation (6.54) for the conductivity correction, the resonant frequency is 4.0835 GHz and the on-load quality factor is reduced to 1416 at a rough root-mean-square value of 4.1 µm. It can be seen from the above results that the resonant frequency decreases with the decrease of conductivity. This is because the skin depth increases with the decrease of conductivity in coaxial cavity, which increases the effective space of cavity.

6.4.2.3 Experimental Verification of the Effect of Rough Lossy Metal Surface on the Electrical Performance of Resonant Cavity

In order to find out the influence of the rough lossy metal surface on coaxial cavity filter, we designed three kinds of rough metal surface, each of which is divided into four kinds of roughness. The metal base plates with different coatings and different roughnesses are shown in Figure 6.17.

Figure 6.18 shows the resonant frequency and S21 test curves of the resonant cavity filter for different roughnesses. From the analysis of Figure 6.18, we can see that for the

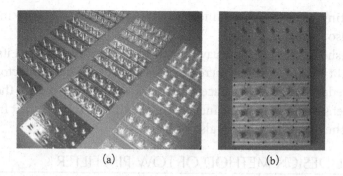

(a) (b)

FIGURE 6.17   Metal substrates with different coatings and roughness. (a) Overall base plate. (b) Comparison of two kinds of gold-plated bottom plates with different roughnesses.

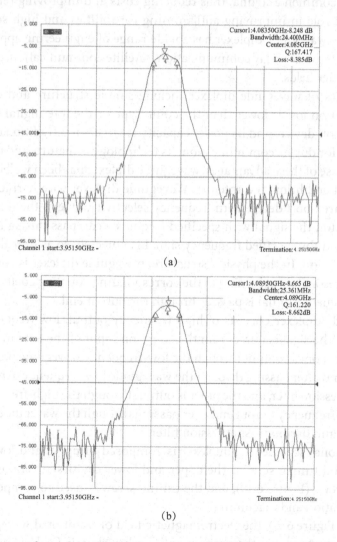

(a)

(b)

FIGURE 6.18   Resonant frequency of coaxial resonant cavity and the test curve of S21. (a) Roughness Ra = 4.12 μm; resonant frequency is 4.0835 GHz. (b) Roughness Ra = 8.03 μm; resonant frequency is 4.0895 GHz.

same metal coating, as the roughness increases, the resonance frequency increases, and the insertion loss also increases.

The analysis shows that the lossy cavity wall will have an impact on its resonance frequency and quality factor; with the increase of loss, the cavity quality factor and resonance frequency are reduced, and the surface roughness will further reduce the quality factor. For the same metal coating, with the increase of roughness, the resonant frequency will be increased, and the insertion loss will also be increased.

6.5 OPTIMAL DESIGN METHOD OF LOW-PIM FILTER

Using waveguide diplexer, signals of different frequency bands can be transmitted and received by the same antenna. Based on this method, radar and communication technology can share a common antenna, thus reducing costs and improving reliability. It has a more prominent role in improving antijamming capabilities and high security requirements. Therefore, waveguide diplexer has a wide range of engineering applications, which can be used in microwave relay communication, satellite communication, microwave measurement and other fields.

The advantages of waveguide diplexer include simple structure and convenient processing; its insertion loss is lower than the comb type or the interdigital type microwave frequency selection device, and its power capacity is larger; as a traditional frequency distribution device for duplex communication, its technology is mature, and its design is relatively easy. Because of these advantages, waveguide diplexer has been widely used in the RF transceiver front end below the antenna. Waveguide diplexer with harmonic suppression capability is a three-port device with frequency selectivity. Due to the frequency selectivity of the resonator, the signal with specified frequency can pass through the device, and the energy beyond the specified frequency signal is reflected, so as to realize the function of frequency selection. In the physical structure, waveguide diplexer is composed of some different single resonators according to the corresponding coupling coefficient, and finally the specified frequency signal is passed through the output end.

The waveguide diplexer consists of high frequency output, low frequency output and common input. The signal source from the common input of the antenna port is divided into two channels; one is output through the low-frequency waveguide port at the low-frequency output end after passing through the waveguide low-frequency band-pass filter and harmonic suppression filter, and the other is output through the high-frequency waveguide port at the high-frequency output end after passing through the waveguide high-frequency band-pass filter and harmonic suppression filter.

In the traditional scheme, the duplexer is composed of upper and lower cavities, fastening screws and tuning screws. The upper and lower cavities are connected as a whole by fastening screws (Figure 6.19), and the tuning screw is used to compensate the errors caused by simulation and machining.

As shown in Figure 6.20, the electromagnetic field of traditional waveguide diplexer is analyzed. It can be seen that the position with higher electric field intensity in the whole cavity is at the tuning screw. The current vector distribution is shown in Figure 6.21. It can be seen that the position with strong current is also at the tuning screw. The tuning screw

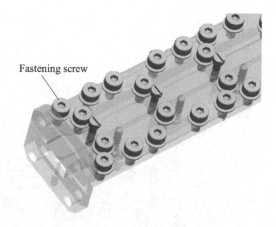

Fastening screw

FIGURE 6.19   Traditional diplexer design.

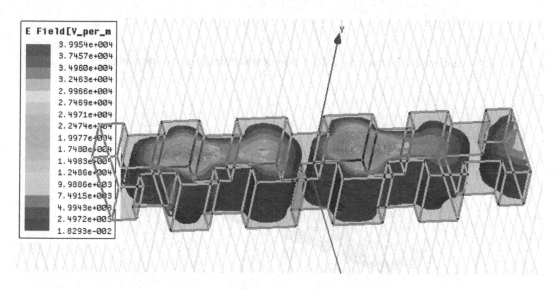

FIGURE 6.20   Amplitude distribution of electric field intensity of traditional waveguide diplexer.

is directly connected with the duplexer cavity through thread with clearance fit, where it is easy to produce nonlinear phenomenon and PIM products under high power due to pollution, unsmooth surface, metal debris and so on.

6.5.1 Low PIM Optimization Design of Waveguide Diplexer

Based on the analysis, when the diaphragm of the waveguide diplexer is moved from the center to the upper and lower edges of the cavity, the tuning screw will also change to the position where the current is relatively weak, so as to reduce the nonlinear current at the tuning screw of the waveguide diplexer, as shown in Figure 6.22.

According to the previous analysis, the tuning screw is directly connected with the duplexer cavity through thread with clearance fit, where it is easy to produce nonlinear phenomenon and PIM products under high power due to pollution, surface roughness,

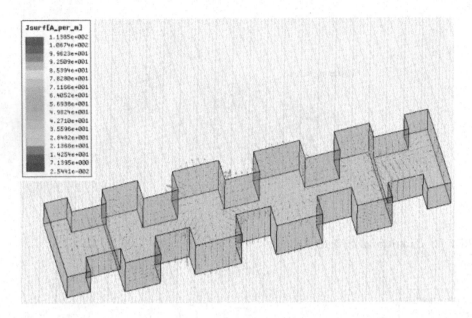

FIGURE 6.21   Surface current vector distribution of traditional waveguide diplexer.

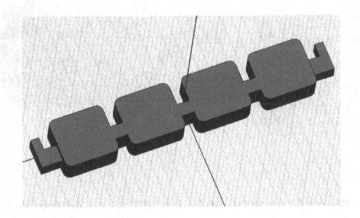

FIGURE 6.22   Low PIM optimization design scheme of waveguide diplexer.

metal debris, etc. If the tuning screw is removed, the PIM level can be greatly reduced as these effects can be eliminated. Therefore, it is necessary to carry out accurate electromagnetic field simulation design, and put forward appropriate and reasonable requirements for machining accuracy in order to ensure the electrical performance of waveguide diplexer. Based on these analyses, this chapter proposes an integrated electroforming processing scheme, as shown in Figure 6.23.

6.5.2 Optimization Design of Low PIM Coaxial Filter

In the design of the experimental sample, the PIM in one position should be highlighted so as to avoid the PIM in other positions. In this way, the surface coating treatment and

FIGURE 6.23   Schematic diagram of low PIM integrated design scheme for waveguide diplexer.

research can be carried out on this position. In the coaxial filter, the current density at the junction of the resonant rod and the cavity bottom surface is relatively large, so the bottom surface of one of the resonant rods is processed, and the other resonant rods and the cavity are integrated to be processed as a whole.

The coupling between resonators is realized by opening a window. The larger the window is, the larger the coupling is. The coupling size of each resonator can be calculated by special software, and the calculated coupling value can be calculated by the formula. The input and output feed is mainly realized by the core of coaxial connector and the feed pole. The closer the feed pole is to the resonator, the greater the feed coupling. There are two ways to connect the core of coaxial connector and the feeder pole: isolation type and welding type. Next, the influence of the two feed modes on PIM is introduced, and the filter with isolated feed mode is used to study the influence of coating on PIM. In order to obtain the position relationship between the resonant rod and the feed pole, the input-output coupling of the filter should be calculated whether it is isolated feed or welded feed.

The three-dimensional model of the coaxial filter in the simulation software HFSS is shown in Figure 6.24. Its working frequency is 2.16 ~ 2.21 GHz, and the in-band insertion loss is less than 0.3 dB. The S-parameter test curve is shown in Figure 6.25.

In order to ensure the basic uniqueness of the PIM source, the isolated feeding third-order coaxial filter is studied, in which one of the resonators is designed separately and fixed at the bottom of the cavity with screws, and the other two resonators are integrated with the cavity. The PIM of this filter and another filter processed integrally with three resonant rods and cavity are tested, and the test data are shown in Figure 6.26.

It can be seen from Figure 6.26 that the PIM of the two filters is quite different, which verifies that this is the main position where PIM occurs and can be plated to study the influence of different coatings on PIM. The bottom of the resonant rod is treated by two processes, oxidation and traditional silver plating, and then the PIM test is carried out

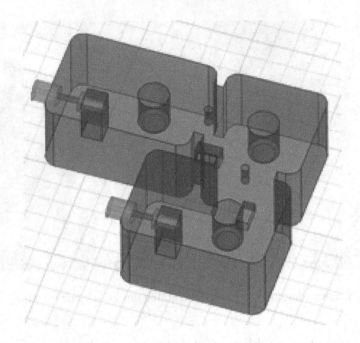

FIGURE 6.24   3D model of low PIM coaxial filter.

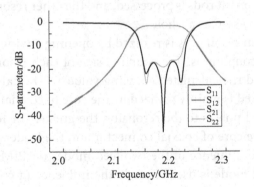

FIGURE 6.25   S-parameter test curve.

under the same environment, and the PIM comparison between the two is obtained, as shown in Figure 6.27.

As can be seen from Figure 6.27, the PIM performance of the filter with surface oxidation treatment is poor. In the actual coating surface treatment process, the coating should be protected to avoid oxide layer.

When the resonant rod is fixed on the bottom of the cavity in time, the PIM of coaxial filter can be effectively suppressed. Therefore, the influence of metal tuning screw on filter PIM is studied, and a new medium tuning screw is designed to replace the metal tuning screw, by which the effectiveness of medium screw on PIM suppression is verified. Figure 6.28 shows the medium tuning screw designed in this book. PIM test is performed on the medium tuning screw filter, as shown in Figure 6.29.

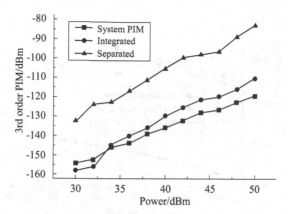

FIGURE 6.26   PIM comparison of two types of filters.

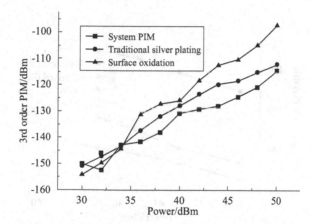

FIGURE 6.27   PIM comparison of surface oxidation and conventional silver plating.

FIGURE 6.28   Medium tuning screw.

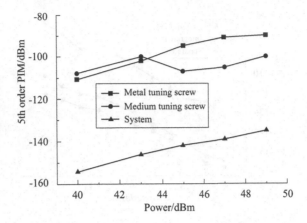

FIGURE 6.29   Comparison of PIM curves of the filter with metal tuning screw and medium tuning screw.

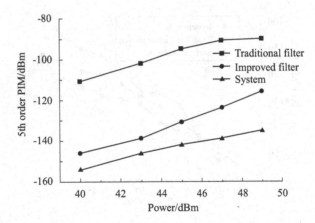

FIGURE 6.30   Comparison of PIM results of improved coaxial filter and traditional coaxial filter.

When the power is more than 45 dBm, the PIM of the filter with metal tuning screw is about 10 dB larger than that of the filter with medium tuning screw. However, the filter with medium tuning screw needs to be precisely designed, and the metal tuning screw can be used for simple microwave components.

According to the above analysis, the main factors affecting PIM are surface coating and tuning screw. In this section, on the basis of optimizing the surface coating and tuning screw, the input and output feed structure of the filter is improved.

PIM test is carried out for the two coaxial filters, and the test curve is shown in Figure 6.30. The PIM of these two kinds of coaxial filters are the 5th order, and the test power is 40~49 dBm. The experimental results show that compared with the traditional PIM design method, this method can effectively reduce 20 dB.

The inner conductor and the feed pole of the coaxial connector are separated by a medium, which can reduce the part of metal welding, so that the PIM of the filter can be effectively reduced. In other microwave components, similar methods can be used to achieve low PIM, which has a good guiding significance for the design of the whole low PIM microwave components.

BIBLIOGRAPHY

1. Sombrin J., Soubercaze - Pun G., Alberti I. Relaxation of the multicarrier PIM specifications of antennas. *The 8th European Conference on Antennas and Propag. Eucap*, 2014: 1647–1650.
2. Gradshteyn I. S., Ryzhik I. M. *Table of Integrals, Series, and Products*. Salt Lake City: Academic Press, 2014.
3. Hoeber C. F., Pollard D. L., Nicholas R. R. PIM product generation in high power communications satellites. Presented At *The AIAA 11th Commutations Satellite Systems Conference*, Mar. 1986, 361–374.
4. Young C. E. The danger of intermodulation generation by RF connector hardware containing ferromagnetic materials. *National Electronic Packaging Conference*, USA. 1975: 20–21.
5. Vicente C., Hartnagel H. L. PIM analysis between rough rectangular waveguide flanges. *IEEE Transactions on Microwave Theory and Techniques*, 2005, 53 (8): 2515–2525.
6. Wang Q., Di X.-F., Li Q.-Q., Cui W.-Z. A design of low-passive intermodulation coaxial filter in s-band. *Space Electrical Technology*, 2017, 14(06):49–53.
7. Hall W. J., Gibson M. H., Kunes M. A., et al. The control of PIM products in spacecraft antennas. *IEE Colloquium on PIM Products in Antennas and Related Structures*, London, UK, 1989: 2/1–2/6.
8. Cavalier M., Shea D. Single feed solution for simultaneous x - band and Ka - band satellite communications. *IEEE Milcom 2004 Military Communications Conference*, Monterey, CA, 2004: 160–162.
9. Cheng Z. *A Study on Passive Intermodulation Interference of Base Station Antenna*. Shanghai: Fudan University, 2013.
10. Zhang Q. S., Gong J. G., Xu Z., et al. A low PIM frequency and polarization multiplexing Ku - band feed chains for satellite antennas. *2017 Sixth Asia - Pacific Conference on Antennas and Propagation (APCAP)*, Xi'An, Shaanxi, 2017: 1–3.
11 Clency L. Y. Compact high performance reflector antenna feeds for space application. *AIAA International Communications Satellite Systems Conference* 2010: 1–6.
12. Mark C. Feed for simultaneous X - band and Ka - band operations on large aperture antennas. *Military Communications Conference IEEE*, 2007: 1–5.
13. Patel K. N., Patenaude Y. L - band integrated feed array design for mobile communication satellite. *Symposium on Antenna Technology & Applied Electromagnetics IEEE*, 1990: 64–69.
14. Kunes M., Gibson M., Connor G., et al. Low pim feed chain design techniques for satellite transmit/receive antennas at L And Ku band. *European Microwave Conference IEEE*, 1989: 795–801.
15. Schennum G. H., Rosati G. Minimizing PIM product generation in high power satellites. *IEEE Aerospace Applications Conference IEEE*, 1996: 155–164.
16. Sanford J. PIM considerations in antenna design. *Antennas & Propagation Society International Symposium IEEE*, 1993: 1651–1654.
17. Brown A. K. PIM products in antennas - an overview. *IEEE Colloquium on PIM Products in antennas and related structures*, 1989: 1/1–1/3.
18. Lui P. L., Rawlins A. D. Passive non - linearities in antenna systems. *IEEE Colloquium on PIM Products in Antennas & Related Structures IET*, 1981: 6/1–6/7.
19. Zhao P., Study Of Passive Intermodulation Interference In UHF Band Of Wireless Communication Systems. Beijing University of Post and Telecommunications.
20. Wang J., Study Of Several Key Problem Of Base Station Array Antenna For Mobile Communication. Qingdao: Shandong University of Science and Technology, 2003.

21. Wang Q., Cui W.-Z., Analysis of low passive intermodulation filter used inspacecraft. *Chinese Space Science and Technology*, 2020, 40(3): 8–12.
22. Rabindra R. S., Eric H. PIM risk assessment and mitigation in communications satellites. *AIAA International Communications Satellite Systems Conference & Exhibit*, 2004: 1–17.
23 Tian L. *Digital Suppression Technique of Passive Intermodulation Interference for Satellite Systems*. Beijing: Beijing Institute of Technology, 2017.

Printed in the United States
by Baker & Taylor Publisher Services